COMPUTATIONAL APPROACHES FOR THE PREDICTION OF pK$_a$ VALUES

QSAR in Environmental and Health Sciences

Series Editor

James Devillers

CTIS-Centre de Traitement de
l'Information Scientifique
Rillieux La Pape, France

Aims & Scope

The aim of the book series is to publish cutting-edge research and the latest developments in QSAR modeling applied to environmental and health issues. Its aim is also to publish routinely used QSAR methodologies to provide newcomers to the field with a basic grounding in the correct use of these computer tools. The series is of primary interest to those whose research or professional activity is directly concerned with the development and application of SAR and QSAR models in toxicology and ecotoxicology. It is also intended to provide the graduate and postgraduate students with clear and accessible books covering the different aspects of QSARs.

Published Titles

Endocrine Disruption Modeling, *James Devillers,* 2009

Three Dimensional QSAR: Applications in Pharmacology and Toxicology, *Jean Pierre Doucet and Annick Panaye,* 2010

Juvenile Hormones and Juvenoids: Modeling Biological Effects and Environmental Fate, *James Devillers,* 2013

Computational Approaches for the Prediction of pKa Values, *George C. Shields and Paul G. Seybold,* 2014

COMPUTATIONAL APPROACHES FOR THE PREDICTION OF pK$_a$ VALUES

George C. Shields
Paul G. Seybold

CRC Press
Taylor & Francis Group
Boca Raton London New York

CRC Press is an imprint of the
Taylor & Francis Group, an **informa** business

CRC Press
Taylor & Francis Group
6000 Broken Sound Parkway NW, Suite 300
Boca Raton, FL 33487-2742

First issued in paperback 2017

ISBN-13: 978-1-4665-0878-1 (hbk)
ISBN-13: 978-1-138-19991-0 (pbk)

Library of Congress Cataloging-in-Publication Data

Shields, George C. (George Charles), 1959- author.
  Computational approaches for the prediction of pKa values / George C. Shields, Paul G. Seybold.
       pages cm. -- (QSAR in environmental and health sciences ; 4)
  Includes bibliographical references and index.
  ISBN 978-1-4665-0878-1 (hardback)
  1. Acids--Basicity. 2. Dissociation. 3. Chemistry, Physical and theoretical. I. Seybold, Paul G., author. II. Title.

QD477.S54 2014
546'.24--dc23   2013036872

Visit the Taylor & Francis Web site at
http://www.taylorandfrancis.com

and the CRC Press Web site at
http://www.crcpress.com

Dedication

To my parents and to my Brady Bunch family, GS1-6 (GCS)

To my students and my wife (PS)

Contents

Series Introduction

The correlation between the toxicity of molecules and their physicochemical properties can be traced to the 19th century. Indeed, in a French thesis entitled "Action de l'alcool amylique sur l'organisme" (Action of amyl alcohol on the body), which was presented in 1863 by A. Cros before the Faculty of Medicine at the University of Strasbourg, an empirical relationship was made between the toxicity of alcohols, their number of carbon atoms as well as their solubility. In 1875, Dujardin-Beaumetz and Audigé were the first to stress the mathematical character of the relationship between the toxicity of alcohols and their chain length and molecular weight. In 1899, Hans Horst Meyer and Fritz Baum, at the University of Marburg showed that narcosis or hypnotic activity was in fact linked to the affinity of substances to water and lipid sites within the organism. At the same time at the University of Zurich, Ernest Overton came to the same conclusion providing the foundation of the lipoid theory of narcosis. The next important step was made in the 1930s by Lazarev in St. Petersburg who first demonstrated that different physiological and toxicological effects of molecules were correlated with their oil-water partition coefficient through formal mathematical equations in the form $\log C = \log P_{oil/water} + b$. Thus, the Quantitative Structure-Activity Relationship (QSAR) discipline was born. Its foundations were definitively fixed in the early 1960s by the seminal works contributed by C. Hansch and T. Fujita. Since that period, the discipline has gained tremendous interest and now the QSAR models represent key tools in the development of drugs as well as in the hazard assessment of chemicals. The new REACH (Registration, Evaluation, Authorization, and Restriction of Chemicals) legislation on substances, which recommends the use of QSARs and other alternative approaches instead of laboratory tests on vertebrates, clearly reveals that this discipline is now well established and is an accepted practice in regulatory systems.

In 1993, the journal *SAR and QSAR in Environmental Research* was launched by Gordon and Breach to focus on all the important works published in the field and to provide an international forum for the rapid publication of SAR (Structure-Activity Relationship) and QSAR models in (eco) toxicology, agrochemistry, and pharmacology. Today, the journal, which is now owned by Taylor & Francis and publishes three times more issues per year, continues to promote research in the QSAR field by favoring the publication of new molecular descriptors, statistical techniques, and original SAR and QSAR models. This field continues to grow very rapidly and many subject areas that require larger developments are unsuitable for publication in a journal due to space limitation.

This prompted us to develop a series of books entitled "QSAR in Environmental and Health Sciences" to act in synergy with the journal. I am extremely grateful to Colin Bulpitt and Fiona Macdonald for their enthusiasm and invaluable help making the project become a reality.

This fourth book in the series clearly describes how computational methods can successfully be used to estimate the pK_a values of structurally diverse molecules for a better understanding of their physicochemical behavior.

At the time of going to press, two books are in the pipeline. One deals with reproductive and developmental toxicology while the other one discusses the long history of the topological indices in QSAR and QSPR and shows the new developments in the field.

I gratefully acknowledge Hilary Rowe for her willingness to assist me in the development of this series.

James Devillers

Series Editor

CTIS—Centre de Traitement de l'Information Scientifique

Rillieux-La-Pape, France

Foreword

The phenomenon of acidity and the nature of the pH scale are concepts encountered remarkably early on in the study of chemistry. What first science course does not at some point involve dipping pieces of pH paper into various common substances to observe their relative acidities or basicities? Perhaps this prominence in science curricula is motivated at least in part by how important a role pH plays in so many processes, and especially in biology, where remarkably small changes in blood pH, for example, can have devastating consequences for an organism.

Interestingly, by the time one reaches somewhat more advanced levels of study, one is likely to encounter a pH meter that is calibrated to be accurate to hundreds, perhaps even thousands, of a pH unit, and certainly one can look up long lists of experimental acid dissociation constants, typically in the form of pK_a values, that have been measured to exquisite accuracy.

This is a theorist's nightmare.

At room temperature, a change in 1/10 of a log unit in a pK_a value implies a change of 0.14 kcal/mol in the corresponding acid dissociation constant. The phrase "chemical accuracy" has come, for theorists, to imply accuracy to within 1.0 kcal/mol. Thus, application of a model expected to offer chemical accuracy would imply an expected error in a pK_a of roughly ± 0.7 log units—one certainly would not want to experience such a variation in pH in one's personal bloodstream (although one would not be burdened with said experience for overly long).

Moreover, for pK_a values, it is extremely difficult to achieve even this rather modest goal of chemical accuracy. Why? The trouble is that from a computational standpoint, one typically imagines a Born-Haber cycle approach to predicting the free energy change associated with acid ionization. Thus, one computes first the gas-phase process, where perhaps chemical accuracy is achievable, although the gas-phase deprotonation free energy is certain to be a number typically on the order of hundreds of kcal/mol. Then, one adds to this the solvation free energies of the typically ionic products, including the proton (or lyonium cation, if one prefers, as this book explains in great detail)—these again have magnitudes on the order of negative hundreds of kcal/mol. And, finally one subtracts the free energy of solvation of the conjugate acid reactant. Thus, the final dissociation free energy in solution is typically one or two orders of magnitude smaller than the large components that largely cancel one another when added together, and the opportunities for error propagation are enormous!

Interestingly, the chemical literature is replete with individual research papers claiming great accuracy with one protocol or another for the prediction of pK$_a$ values. I consider few, if any, to be terribly honest. Over a small range of chemical functionality, there will often be some combination of one particular gas-phase protocol with one particular solvation protocol that will give good results over that range, and reports of such limited success stories tend to be the fodder of most published papers. However, achieving high accuracy with a single approach for an *arbitrary* acid still presents extraordinary challenges.

In this text, Shields and Seybold carefully review the thermochemical principles involved in the measurement and prediction of acid dissociation constants. They further provide an extraordinarily useful comparison of several theoretical models that have been proposed, whether for more specific or more general applications, examining their relative utilities and ranges of applicability. Importantly, the authors tightly couple their analysis to the rich physical organic chemistry associated with all of the various functional groups that may serve as conjugate acids or bases, and they do so not only for the case of aqueous solution, but also for non-aqueous solvents. They wrap up with some welcome discussion of temperature effects, isotope effects, and other phenomena that can extend the utility of acid dissociation constants as reporters of electronic structure and reactivity.

This book will be a critical resource not only for those who want to predict an unknown pK$_a$ value, but also for those who want to understand *why* a given compound has a particular pK$_a$ value, and how that value places it into a more general physical organic context. Certainly, I imagine that most copies will be dog-eared and thumb-worn early in their lifetimes; I know mine will.

Christopher J. Cramer
University of Minnesota

Acknowledgments

My grandparents, and my grandfathers in particular, were influential in shaping my worldview. Charles Shields, an electrical engineer holding many patents, nurtured my early interests in science. George Vetter inspired me to do my very best to make a difference with my life. My parents, Robert and Margaret Shields, raised five children who are all making a difference. My high school chemistry teacher, Mary Ann Baner (who still sends me homemade cookies), was the first person to make chemistry cool. As a student at Georgia Tech I owe a debt of gratitude to many people, but in particular the late Peter Sherry, Bud Suddath, Bob Pierotti, and Bill Landis; and to Aaron Bertrand, Sally Hammock, and Charlie Liotta. I owe a huge debt of gratitude to my graduate research advisor, Tom Moran, who taught me how to think like a scientist. I thank my postdoctoral mentor at Yale, Tom Steitz, who taught me that all problems in science are hard to solve, so you might as well try to solve the most difficult ones. I owe a special thanks to Steve Schultz and Anna DiGabriele, who were great to work with and are wonderful friends, and helped me decide on my life's work as a teacher-scholar at undergraduate institutions. We had a healthy invasion of British postdocs when I was at Yale in the late 1980s, and in particular Adrian Goldman patiently taught me protein x-ray crystallography. At Lake Forest College I was lucky to work with colleagues such as Mike Kash, Laura Kately, Bill Martin, and Lee Thompson; Bill's words of wisdom in particular guide me to this day. I worked with 18 undergraduate students on research projects at Lake Forest, and Karl Kirschner, Mariela Serrano, Ed Sherer, Gordon Turner, and Rebecca Turner have all been instrumental in the pK_a work that is the subject of this book. Ed and Gordon were there at the beginning. Ed developed the initial direction in collaboration with Don Landry at the Columbia University College of Physicians and Surgeons, and Karl helped immensely when he rejoined my lab as a senior scientist at Hamilton College. Don Landry was the single most important inspirational source of our work on pK_a, when he asked me one day in 1996 if it was possible to calculate the pK_a values of some hapten molecules he wanted to synthesize, and I, in ignorance of the difficulty of such a task, or perhaps mindful of Tom Steitz's advice, said that we would be willing to try. The American Chemical Society/ Petroleum Research Fund (ACS/PRF), NIH (AREA), and NSF (RUI) all supported this work from 1998 to 2008, and allowed my lab to work with

72 undergraduates, three senior scientists, and a postdoctoral associate during this time. Sarah Tschampel and Annie Toth were my first Hamilton students to work on this project, and Matt Liptak was the key student who helped us understand that it was possible to predict reasonably accurate pK$_a$ values using modern methods, which we began publishing in 2001. Meghan Dunn, Matroner George, Daniel Griffith, Beth Hayes, Christy House, Katrina Lexa, Brent Matteson, James McConnell, Matt Palascak, Danielle Phillips, Frank Pickard, Emma Pokon, Jaime Skiba, high school student Skylar Ferrara, and postdoc Steve Feldgus all made important contributions to this work. Kristin Alongi's 2008 senior thesis, "Theoretical Calculations of Acid Dissociation Constants," was the basis for the review we wrote and published in the *Annual Reports in Computational Chemistry* in 2010, and which eventually led to this book. I thank my colleagues at Hamilton for their steadfast support of our efforts to make Hamilton a leader in undergraduate research, in particular Karen Brewer, Tim Elgren, Robin Kinnel, and Ian Rosenstein; the then-administrators who steadfastly supported our work: Bobby Fong, David Paris, and Gene Tobin; as well as my great friends Dave Thompson and Betsy Stewart. My work was supported at Armstrong Atlantic University by many colleagues, most notably Steve Jodis, Will Lynch, Delana Nivens, and Ellen Whitford. At Bucknell University, I have had great support from Lisa Aversa, John Bravman, Dee Casteel, Charlie Clapp, Abe Feuerstein, Renée Gosson, Mick Smyer, Tim Strein, Berhane Temelso, Ann Tlusty, and many others. I thank my colleagues and collaborators at research institutions, especially Igor Alabugin, Tom Anderson, Jim Bennett, Chris Cramer, Ted Dibble, Ken Jordan, Karl Kirschner, Charlie Laughton, John Lazo, Modesto Orozco, Kennie Merz, Vince Ortiz, Brooks Pate, Adrian Roitberg, Ed Sherer, David Sherrill, Greg Tschumper, and Veronica Vaida, all of whom supported and encouraged our work. I thank all of my MERCURY undergraduate faculty colleagues; all of you make a difference in the work you do. I am very fortunate to have a great coauthor, Paul Seybold, who I first met in 1995 when I was just starting this work. His friendship and support over the past two decades have been steadfast and motivating. I thank all of my children, Grace, Elaine, Luke, Crystal, Kechi, and Nneka, who have supported their scientist father throughout the years. Finally, I thank my favorite writer, my best friend, and the one who keeps life exciting, CoCo Harris.

George C. Shields

I thank all of the students at Wright State University and our wonderful collaborators at other universities who have cooperated on many of the studies described in this book, as well as my wife for her patience during the period during which this book was written. I thank my colleague and friend George Shields for many years of pleasant and fruitful conversations and interactions. I also thank Charlie Perrin at the University of California, San Diego, for a number of helpful discussions related to this work.

Paul Seybold

We thank Professor James Devillers for inviting us to write this book for the QSAR series, and for his thoughtful comments on the manuscript. We thank our editors at Taylor & Francis, Hilary Rowe, acquiring editor, Laurie Schlags, project coordinator, and Amy Rodriguez, project editor, for making the production of this book easy and enjoyable for the authors.

We thank Berhane Temelso for developing the initial designs for the cover art.

GCS & PS

About the Authors

George Shields grew up in Marcellus, New York. He completed his bachelor's, master's, and doctoral degrees in chemistry at Georgia Tech. Although he enrolled in the master's program with the intention of becoming a high school science teacher, he later engaged in doctoral studies in chemical physics, using mass spectrometers to study ion-molecule collisions. In the late 1980s, he became a postdoctoral associate at Yale, working with Tom Steitz (2009 Nobel Laureate in Chemistry), and with Steve Schultz to solve the structure of the catabolite gene activator protein complexed with DNA. After nearly a decade as a faculty member, and in latter years as chair at Lake Forest College, he moved to Hamilton College to chair the department and soon became the Winslow Professor of Chemistry. While at Hamilton he founded the MERCURY (Molecular Education and Research Consortium in Undergraduate Computational Chemistry) consortium to support faculty working with undergraduates. He became the founding dean of the College of Science and Technology at Armstrong Atlantic State University. He is currently a professor of chemistry and dean of the College of Arts and Sciences at Bucknell University. His research uses computational chemistry to investigate atmospheric and biological chemistry.

Paul Seybold grew up in Collingswood, in southern New Jersey where he attended public schools. He received his bachelor's in engineering physics *with distinction* from Cornell University and a PhD in biophysics from Harvard University, working with Martin Gouterman. Following his PhD he did postdoctoral work with Prof. Per-Olov Löwdin at the Quantum Chemistry Group, University of Uppsala, Sweden, and with Prof. Inga Fischer-Hjalmars at the Institute of Theoretical Physics, Stockholm University. He then did a postdoctoral year with Prof. Gregorio Weber in biochemistry at the University of Illinois. Since that time he has been a faculty member and department chair (1999–2004) in the Department of Chemistry at Wright State University in Ohio. He has been a visiting scholar and visiting professor at a number of universities in the United States and Europe. His research interests center on chemical and biochemical applications of quantum chemistry, molecular structure-activity relationships, luminescence spectroscopy, and cellular automata models of complex systems.

Preface

In 1934 Saxton and Meier [1] published a report in which they described measurements of the acid dissociation constants of benzoic acid and its three monochloro derivatives. The results showed an interesting pattern: all of the chloro-derivatives were more acidic than the parent compound, with the *ortho*-substituted compound most acidic and the meta-chloro derivative least acidic. However, the authors closed their report with the lament that "at present, there is no adequate theoretical method of calculating the ionization of a weak electrolyte from its structure."

A great deal has happened in the ensuing eight decades. We now have a fairly good picture of the salient features—electronic, steric, and environmental—that influence a molecule's acidity, and based on these insights, a variety of computational methods have been developed to estimate a compound's acidity from its molecular structure. The most advanced theoretical techniques, although still imperfect (as are many of the measurements, especially at the extremes of the acidity scale) are capable of estimating pK_as to within about ±0.20 of a pK_a unit in many cases.

GCS's introduction to this field started in the mid 1990s, when his group was studying ester hydrolysis and thinking about ways to develop catalytic antibodies against cocaine [2,3]. The Shields' group's collaboration with Don Landry (Columbia University College of Physicians and Surgeons) led to several papers [4–6], but it was Don's question of whether it was possible to calculate the charge state of a hapten before it was synthesized that led to a fruitful search for ways to calculate absolute pK_a values [7–13]. Since 2001 the field of absolute pK_a calculations has exploded [14–16].

PS's introduction to this field came as he was working with an exceptional M.S. graduate student, Kevin Gross, on a quantum chemical study of the changes in the properties of aniline that occur with substitution on the phenyl ring. Kevin added data on the pK_as of the substituted compounds to the property mix and found significant correlations with other properties, such as bond lengths, charges, and energies [17]. We said, "Hmm ... that's interesting," and began looking at additional chemical systems [18,19]. Chad Hollingsworth applied these techniques to benzoic acids [20], and Dr. Warren Kreye examined biophenols [21] and azoles [22]. Several students from a graduate quantum course added their talents [23]. We also had the good fortune to tie up with Chris Hadad at Ohio State University [24,25], Jane Murray and Peter Politzer and their associates at the University of New Orleans [26,27], and Matthew Liptak, Steve Feldgus, and George Shields [11], then at Hamilton College, in later studies.

Our aims in writing this book are twofold. First, we want to describe the insights that have been gained about the intrinsic and extrinsic features that influence a molecule's acidity. And second, we want to discuss, in a critical way, the strengths and weaknesses of the theoretical techniques that have been developed to exploit these insights. We also want to show how these techniques have been employed to gain information about the acidities of different classes of chemical compounds. Because of the enormous scope of the literature in this field, we make no claim of completeness. Rather, we have attempted to provide general discussions and representative references that may be employed to further explore different individual subjects.

We are thankful for our students and colleagues around the world who have played central roles in measuring acidities, developing new theoretical techniques, and using them in exploring this fascinating field.

1 Introduction

Acid dissociation constants are essential for understanding many fundamental reactions in chemistry and biochemistry [15,28]. pK_a values are a convenient way to specify the dissociation constants for weakly acidic or basic groups, and thus are extremely informative [29]. When comparing compounds, pK_a values allow scientists to compare acid strengths, base strengths, Gibbs free energy changes, and equilibrium constants of ionization reactions. In an acid-base equilibrium reaction, the pK_a allows an easy prediction of the favored direction for that equilibrium as well as the concentrations of the individual species at a given pH. For nucleophilic substitution reactions, the pK_a values allow prediction of the relevant strength of the nucleophile and the best leaving groups [28]. For the acid dissociation reaction:

$$HA_{(aq)} \rightleftharpoons A^-_{(aq)} + H^+_{(aq)} \tag{1.1}$$

the pK_a of the acid HA is then defined as

$$pK_a = -\log K_a \tag{1.2}$$

$$K_a = [A^-_{(aq)}]\,[H^+_{(aq)}]/[HA_{(aq)}] \tag{1.3}$$

where K_a is the equilibrium constant for the dissociation (reaction 1.1). Of course the thermodynamic definition of any equilibrium constant is in terms of activities, and activities are the products of activity coefficients and concentrations [30]. In addition, because the concentration of water is constant in all but the most concentrated solutions, K_a in equation 1.3 includes the concentration of water (around 55 M, varying with temperature) [30,31].

Note that reaction 1.1 represents acid dissociation in water, or aqueous solution, yet pK_a values can in principal be determined in any solvent. Deprotonation reactions such as reaction 1.1 are most often written with (aq) representing the unknown micro and bulk solvation effects of the solvent, water in this case. The exact nature of the solvated proton in water is unknown at this time and is the subject of much research; the standard practice is to use a nominal definition for the solvated proton, and to represent the solvated proton as $H^+_{(aq)}$ [32,33]. This practice is justified because the solvated proton exists in a cluster of many water molecules, and proton transfer reactions depend on the magnitude of the proton's solvation free energy and not on the geometric details of the proton's solvation shell [32,34].

One can also describe the equilibrium constant for the protonation of a base, K$_b$, and it is a simple matter to show that pK$_a$ and pK$_b$ are related through the pK$_w$ by the following expression:

$$pK_b = pK_w - pK_a$$

where K$_w$ is the product of the [H$^+$] and [OH$^-$] concentrations [30]. Thus it is a matter of convenience whether to use pK$_a$ or pK$_b$ values, as they are related by a constant. The value of pK$_w$ varies with temperature, and is very close to 14.0 for pure water at 25°C [31].

In many cases pK$_a$ values can be readily measured experimentally, but often chemists are interested in the pK$_a$ values of molecules that have not yet been synthesized or for which experiments are not easy or straightforward. For instance, in a series of molecules that are of pharmaceutical interest, slight variations in substituents will change the pK$_a$ values, and this then impacts the solubility and permeability of the potential drug or pharmaceutical.

Predicting the pK$_a$ in advance saves much time and effort [15,35,36]. Therefore, the ability to computationally calculate pK$_a$ values accurately is important for scientific advancements in biochemistry, medicinal chemistry, and other fields. Chemical accuracy, however, is hard to achieve. Computationally calculating acid dissociation constants is a demanding and arduous process because an error of 1.36 kcal/mol in the change of free energy of reaction 1.1 results in an error of 1 pK$_a$ unit [8,11]. There are numerous studies that use a variety of theoretical methods in attempts to obtain chemical accuracy. In recent years, there have been new developments, but many discrepancies still exist [14,16,37]. The aim of this work is to compare most of the significant methods for accurate pK$_a$ calculations for researchers interested in using the most appropriate method for their system of interest. We focus on three areas of development: absolute pK$_a$ calculations using thermodynamic cycles and gas-phase and solution-phase free energy calculations, relative pK$_a$ calculations that simplify the calculation for systems with a common moiety where the experimental value is known, and correlation methods.

2 Absolute pK$_a$ Calculations[*]

Absolute pK$_a$ calculations rely on four items: (1) a thermodynamic cycle that relates the gas phase to the solution phase; (2) knowledge of relevant experimental values; (3) accurate gas-phase calculations; and (4) accurate solvation calculations. Once an appropriate thermodynamic cycle is devised, then state-of-the-art quantum chemical calculations are used with the cycle to estimate pK$_a$ values [15]. These estimates can be quite accurate for small molecules, but are of course limited to systems small enough where highly accurate quantum mechanical values can be calculated.

2.1 THERMODYNAMIC CYCLES

Numerous thermodynamic cycles have been used to calculate pK$_a$ values [9]. A series of excellent reviews of thermodynamic cycles and the most common solvation models used for pK$_a$ calculations has recently been published by Ho and Coote [14,16,37]. One of the most common methods is depicted in Figure 2.1, based on reaction 1.1 from Chapter 1 [8,9,11].

In Figure 2.1, ΔG_{aq} represents the overall change in Gibbs free energy of this reaction in aqueous solution, ΔG_{gas} is the change in the gas-phase Gibbs free energy, and ΔG_{sol} is the change in Gibbs free energy of solvation. The reaction of interest in this case is the aqueous Gibbs free energy change for deprotonation, which is governed by the equilibrium constant for reaction 1.1 in Chapter 1. Based on the diagram, pK$_a$ is calculated using the following equations:

$$pK_a = \Delta G_{aq}/RT\ln(10) \tag{2.1}$$

$$\Delta G_{aq} = \Delta G_{gas} + \Delta\Delta G_{sol} \tag{2.2}$$

where

$$\Delta G_{gas} = G_{gas}(H^+) + G_{gas}(A^-) - G_{gas}(HA) \tag{2.3}$$

---

[*] From Alongi, K.S., and Shields, G.C., Theoretical calculations of acid dissociation constants: A review article, Annual Reports in Computational Chemistry, pp. 113–138, 2010. Copyright © 2010, with permission from Elsevier.

$$\Delta G_{gas}$$

$$HA_{(g)} \quad \rightarrow \quad A^-_{(g)} \quad + \quad H^+_{(g)}$$

$$\uparrow -\Delta G_{sol}(HA) \quad \downarrow \Delta G_{sol}(A^-) \quad \downarrow \Delta G_{sol}(H^+)$$

$$HA_{(aq)} \quad \rightarrow \quad A^-_{(aq)} \quad + \quad H^+_{(aq)}$$

$$\Delta G_{aq}$$

FIGURE 2.1   Proton-based thermodynamic cycle.

and

$$\Delta\Delta G_{sol} = \Delta G_{sol}(H^+) + \Delta G_{sol}(A^-) - \Delta G_{sol}(HA) \tag{2.4}$$

All of these Gibbs free energy values can be calculated using quantum chemistry, except $\Delta G_{sol}(H^+)$ and $G_{gas}(H^+)$, which must be determined experimentally or using thermodynamic theory. Recently it has become possible to make theoretical estimates of $\Delta G_{sol}(H^+)$ using a combined explicit-implicit approach, which is discussed in Section 2.11. In the future as computational power increases it should be possible to calculate $\Delta G_{sol}(H^+)$ explicitly. How these various quantities are determined will be discussed in the corresponding sections later in this chapter. A similar thermodynamic cycle that is often used is based on the acid dissociation equation of a protonated acid [38]:

$$HA^+_{(aq)} \rightarrow A_{(aq)} + H^+_{(aq)} \tag{2.5}$$

In this case, reaction 2.5 leads to a thermodynamic cycle similar to that in Figure 2.1, which was based on reaction 1.1. Accordingly, equation 2.3 becomes $\Delta G_{gas} = G_{gas}(H^+) + G_{gas}(A) - G_{gas}(HA^+)$ and equation 2.4 becomes $\Delta\Delta G_{sol} = \Delta G_{sol}(H^+) + \Delta G_{sol}(A) - \Delta G_{sol}(HA^+)$. Additional thermodynamic cycles are based on the acid's reaction with a water molecule, as depicted in the following reactions:

$$HA_{(aq)} + H_2O_{(aq)} \rightarrow A^-_{(aq)} + H_3O^+_{(aq)} \tag{2.6}$$

$$HA^+_{(aq)} + H_2O_{(aq)} \rightarrow A_{(aq)} + H_3O^+_{(aq)} \tag{2.7}$$

$$HA^+_{(aq)} + H_2O_{(aq)} \rightarrow H_2O{\cdot}A_{(aq)} + H^+_{(aq)} \tag{2.8}$$

$$HA_{(aq)} + H_2O_{(aq)} \rightarrow H_2O{\cdot}A^-_{(aq)} + H^+_{(aq)} \tag{2.9}$$

Reactions 2.6–2.9 lead to equivalent thermodynamic cycles similar to Figure 2.1. In these cycles, however, the number of waters included in the cycle may vary [39]. A key point is that the concentrations of all species in a given thermodynamic cycle must have the same standard state, typically 1 M, and when water is introduced into the cycle as a reactant it should also have this standard state [40,41]. One limitation of the cycles derived from reactions 2.6–2.9 is that the hydronium ion's change in Gibbs free energy of solvation is difficult to calculate because of its high charge. However, one can use the accepted experimental $\Delta G_{sol}(H_3O^+)$ value of –110.3 kcal/mol in the 1 M standard state and the accepted experimental value for $\Delta G_{sol}(H_2O)$ of –6.32 kcal/mol [39,42,43]. This reduces the number of computations and ensures similar accuracy as using the proton-based thermodynamic cycle and the most recently accepted experimental value for $\Delta G_{sol}(H^+)$.

Of course, the exact nature of H$^+$(aq), or H$_3$O$^+$(aq), is unknown, as some indeterminate cluster of water molecules necessarily surrounds the proton, so the choice of the entity to use in a written reaction is arbitrary, and as outlined in Chapter 1 the simplest choice is H$^+$(aq) [32,33]. The addition of another water to the aqueous reaction, such as in 2.8 or 2.9, also results in more calculations. This increases computational error in determining pK$_a$, as the total number of manipulated numbers increases from reaction 2.5 to reaction 2.9. For a thorough understanding of the standard state issues that arise when adding water to thermodynamic cycles, study the recent Goddard group paper on this topic [41].

Another thermodynamic cycle often used is derived from the acid dissociation reaction with a hydroxide ion to produce water [38,44]:

$$HA_{(aq)} + OH^-_{(aq)} \rightarrow A^-_{(aq)} + H_2O_{(aq)} \tag{2.10}$$

$$HA^+_{(aq)} + OH^-_{(aq)} \rightarrow A_{(aq)} + H_2O_{(aq)} \tag{2.11}$$

The limitations of these cycles are similar to those used for reactions 2.6–2.9, where the additional reactant increases the number of calculations and, consequently, the computational error. The change in Gibbs free energy of solvation for OH$^-$ is also difficult to determine computationally because OH$^-$ is a diffuse anion with an indeterminate hydration shell, but the experimental value can be used instead. The accepted experimental $\Delta G_{sol}(OH^-)$ value is –104.7 kcal/mol in the 1 M standard state [39,45]. Again, as long as the correct experimental values are used for the Gibbs free energy of solvation of H$^+$, H$_3$O$^+$, H$_2$O, and OH$^-$, these different thermodynamic cycles should yield similar results. Increasing the number of species primarily affects the calculation of ΔG_{gas}, as equation 2.3 now contains another reactant and product. Reactions 2.10 and 2.11 can be modified by hydrating the hydroxide ion with n+m waters, resulting in n waters being attached to A$_{(aq)}$ or A$^-_{(aq)}$ and m waters returning to the solvent H$_2$O$_{(aq)}$ after reaction, in the cluster-continuum approach first outlined by Pliego and Riveros [45]. The difficulty in computing accurate Gibbs free energy of solvation values for anions is discussed later in Section 2.3.

The thermodynamic cycles derived from reactions 2.6–2.11 are depicted in a manner similar to Figure 2.1. The pK$_a$ values are then calculated using equations 2.1 and 2.2; equations 2.3 and 2.4 are modified to correlate to the specific thermodynamic cycle.

2.1.1 LIMITING EXPERIMENTAL VALUES

A proton contains no electrons, and its Gibbs free energy cannot be calculated quantum mechanically. Calculation of this energy using the standard equations of thermodynamics and the Sackur-Tetrode equation [46] yields the same value as can be deduced experimentally from the National Institute of Standards and Technology (NIST) database. The translational energy of 1.5RT combined with PV = RT and H = E + PV yields a value of H°(H$^+$) equal to 5/2(RT) or 1.48 kcal/mol. Use of the Sackur-Tetrode equation yields the entropy, TS(H$^+$) = 7.76 kcal/mol at 298 K and 1 atm pressure. Finally, since G = H − TS, G°(H$^+$) = −6.28 kcal/mol.

One of the main sources of error in pK$_a$ calculations is the value used for the Gibbs free energy of solvation for H$^+$, which is explicitly needed in certain thermodynamic cycles. In contrast to G°(H$^+$), over the past 15 years the experimental value for ΔG_{sol}(H$^+$) has changed considerably, from an accepted range of −254 to −261 kcal/mol in 1991 [44] to the now accepted −265.6 ± 1 kcal/mol [15]. The basic uncertainty in any determination of the Gibbs free energy of solvation of an ion is that ions are never isolated in solution. Determination of, say, ΔG_{sol} for an anion, A$^-$, can be made if the corresponding value for a cation, C$^+$, is already known. That way when the ΔG_{sol} is measured for the salt CA, the value for ΔG_{sol} of A$^-$ can be determined by difference. In practice all ionic solvation values are referenced against the value for H$^+$; every time the value for ΔG_{sol}(H$^+$) changes, the values for all of the other ions in the various databases change as well. The older value for ΔG_{sol}(H$^+$) of −261.4 kcal/mol was estimated from an average of five independent measurements of the hydrogen ion electrode [15,47]. More recent values reflect clever experimental and computational approaches that are discussed below.

When we began our work on pK$_a$ calculations, using carboxylic acids and phenols as test cases, we tried to use the then accepted value of −261.4 kcal/mol. Like previous researchers, it was not possible to calculate accurate absolute pK$_a$ values [47–50]. Karplus and co-workers reported that the continuum dielectric methods were not accurate enough to yield accurate absolute solvation Gibbs free energies [49]. By 2001, new solvation methods had been developed, and relative pK$_a$ values were easily determined, leading to a suspicion that the value for $\Delta G_{sol}(H^+)$ was erroneous [7]. We used an experimental thermodynamic cycle for acetic acid dissociating to H^+ and the acetate anion to derive an experimental value for $\Delta G_{sol}(H^+)$ of –264.61 kcal/mol [8,11]. At that time, this was by far the most negative value used for these types of calculations, and it was a bit shocking to be using a value that was more than 3 kcal/mol lower than that obtained from the hydrogen ion electrode. However, two groups had used a combined explicit-implicit theoretical approach to obtain $\Delta G_{sol}(H^+)$ values of –264.4 and –264.3 kcal/mol for a standard state of 1 M [34,51]. In addition, Coe and co-workers had used experimental ion-water clustering data to derive a value for $\Delta G_{sol}(H^+)$ of –264.0 kcal/mol [42]. At the time, much of the computational chemistry community thought that this value was for a standard state of 1 M [52–54], yet after a few years of confusion it was determined that the correct standard state was 1 atm, so that changing to a standard state of 1 M changed the value of $\Delta G_{sol}(H^+)$ to –265.9 kcal/mol [15,43,55]. In addition, Goddard and co-workers have recently shown that by including concentration corrections to Zhan and Dixon's high-level *ab initio* calculations of the hydration Gibbs free energy of the proton [51], that their value is corrected to –265.63 ± 0.22 kcal/mol [41]. We owe the Camaioni, Goddard, and Cramer/Truhlar research groups a debt of gratitude for bringing clarity to this issue. Discussion of the standard state conversion follows.

The newest value was originally determined by Tissandier et al. in 1998 using correlations between ΔG_{sol} of neutral ion pairs and experimental ion-water clustering data, which is known as the cluster pair approximation method [47]. Kelly, Cramer, and Truhlar confirmed this value in 2006 using a similar method but larger data set [55]. Experimental uncertainty, however, does still exist, which introduces uncertainty in all of these procedures. In 2010 Donald and Williams developed an improved cluster pair correlation method and determined a value for the Gibbs free energy of solvation for the proton at a 1 atm standard state to be –263.4 kcal/mol [56], which corresponds to –265.3 kcal/mol for a standard state of 1 M. We note that the most accurate values of a wide range of unclustered cations and anions, based on the accepted value for the Gibbs free energy of solvation of H^+, are given in reference 55. For clustered ions, standard state corrections for concentration of waters clustered to the ions must be included [41], and the corrected values for clustered ions have been published in the supplemental information of a recent paper on Minnesota solvent models [57].

The absolute solvation Gibbs free energy of a proton can also be calculated using high-level gas phase calculations with a supermolecule-continuum approach, involving a self-consistent reaction field model.

The change in Gibbs free energy of solvation is calculated by adding waters to H$^+$ until a converged value is reached. The solvent is approximated by a dielectric continuum medium that surrounds the solute and any added explicit waters, and the number of quantum mechanically treated explicit solvent molecules is increased to improve the calculations. Using this approach, Zhan and Dixon calculated $\Delta G_{sol}(H^+) = -264.3$ kcal/mol [51]. Correction to the 1 M standard state changes this value to -265.63 ± 0.22 kcal/mol [41]. Thus the most recent experimental and theoretical determinations of $\Delta G_{sol}(H^+)$ are now approximately 0.3 kcal/mol from the -265.6 theoretical determination, lending confidence in using either the -265.3, -265.63, or -265.9 value, and lowering the uncertainty in this value to be closer to 1 kcal/mol. Using these values we estimate that the most correct value for $\Delta G_{sol}(H^+)$ is -265.6 ± 1 kcal/mol. Zhan and Dixon's corrected value is the standard against which all future explicit quantum mechanical calculations of $\Delta G_{sol}(H^+)$ will be evaluated [51].

We have seen that the standard state of $\Delta G_{sol}(H^+)$ must be taken into account to produce reliable results. In this last part of this section we outline the details of standard state conversions.

Gibbs free energies can be calculated using either an ideal gas at 1 atm as a reference for gas phase calculations or with an ideal gas at 1 mol/L; the latter standard state is used in most quantum chemistry programs that calculate the Gibbs free energy of solvation. The values for ΔG_{sol} and ΔG_{gas} depend on which standard state is used in their determination. Furthermore, a homogenous equilibrium, where all of the species are in the same standard state, is necessary to obtain reliable pK$_a$ results. The conversion of the 1 atm standard state to the 1 mol/L standard state can be derived from the relationship between the equilibrium constant based on a concentration of 1 M, K$_c$, and the equilibrium constant expressed in terms of pressure in the 1 atm standard state, K$_p$. The relationship between the two constants is derived for the following general reaction:

$$aA_{(g)} \Leftrightarrow bB_{(g)} \qquad (2.12)$$

The corresponding equilibrium constants are:

$$K_c = [B]^b/[A]^a \qquad (2.13)$$

$$K_p = P^b_B/P^a_A \qquad (2.14)$$

Using the ideal gas law, PV = nRT, where R is given by 0.8206 L·atm/K·mol, we rewrite equation 2.14:

$$K_p = (n_B RT/V)^b/(n_A RT/V)^a = [(n_B/V)^b/(n_A/V)^a] \cdot (RT)^{b-a} \qquad (2.15)$$

Since n_B/V and n_A/V now have the units mol/L, they can be replaced by the concentrations of A and B and simplified in terms of K_c:

$$K_p = \{[B]^b/[A]^a\} \cdot (RT)^{b-a} = K_C(RT)^{\Delta n} \tag{2.16}$$

where Δn is the change in the number of moles, b–a. Equation 2.16 can be used to show the relationship between the Gibbs free energies in different standard states. The relationship between the 1 M state and the standard state of 1 atm is [55]:

$$G^* = G^\circ + \Delta G^{\circ \rightarrow *} \tag{2.17}$$

$$\Delta G^* = \Delta G^\circ - \Delta G^{\circ \rightarrow *} \tag{2.18}$$

where G^* is for the 1 mol/L standard state, G° is for the 1 atm standard state, and $\Delta G^{\circ \rightarrow *}$ represents the conversion from the 1 atm to the 1 mol/L standard state. We can determine $\Delta G^{\circ \rightarrow *}$ by using equation 2.16 and the relations between the equilibrium constants and the free energies at the two different standard states:

$$\Delta G^* = -RT\ln K_c \tag{2.19}$$

$$\Delta G^\circ = -RT\ln K_p \tag{2.20}$$

Using equations 2.16, 2.19, and 2.20, we can find the relationship between the two Gibbs free energies at 298.15 K:

$$\Delta G^* = \Delta G^\circ - RT\ln(RT)^{\Delta n} = \Delta G^\circ - RT\ln(24.4654)^{\Delta n} \tag{2.21}$$

In relation to equation 1.2, this illustrates that for the dissociation reaction 1.1, $AH_{(aq)} \rightarrow A^-_{(aq)} + H^+_{(aq)}$, where $\Delta n = 1$:

$$\Delta G^{\circ \rightarrow *} = RT\ln(24.4654) \tag{2.22}$$

At 298.15 K this conversion of standard states, $\Delta G^{\circ \rightarrow *}$, equals 1.89 kcal/mol. When $\Delta G_{sol}(H^+) = -265.6$, it is in the gas-phase 1 M standard state. If reported with the 1 atm standard it is –263.7 kcal/mol. In calculating accurate pK$_a$ values, one must be aware of the standard state because a difference of 1.89 kcal/mol can cause significant error and unreliable values [8,9,11,55].

The G_{gas} H$^+$ value cannot be determined quantum mechanically. Its value, however, has less uncertainty, and is the same whether determined from experimental values available on the NIST website [10,12,58] or from the Sackur-Tetrode equation [46], and is consistently accepted as –6.28 kcal·mol^{-1} for a standard state of 1 atm [8,11].

2.2   GAS PHASE GIBBS FREE ENERGY CALCULATIONS

The gas phase Gibbs free energy calculation is the smallest source of error in absolute pK$_a$ calculations [15]. High levels of theory, such as CBS-QB3 [59] and CBS-APNO [60], produce reliable ΔG_{gas} values with root-mean-square deviations of 1.1–1.6 kcal/mol from the Gibbs free energy of gas-phase deprotonation reactions complied in the NIST online database [10,13,58]. With today's computers and focusing on small molecules, coupled cluster CCSD(T) calculations extrapolated to the complete basis set limit can give gas phase Gibbs free energies as accurate or even more accurate than experiment. CCSD(T) stands for **C**oupled **C**luster with all **S**ingle and **D**ouble substitutions along with a quasi-perturbative treatment of connected **T**riple excitations, and as of this writing is considered the gold standard in *ab initio* quantum chemistry. Details will be discussed below. Practitioners would like to achieve accurate results without using such a computationally expensive level of theory [12]. Combinations of different methods, such as model chemistries, density functional theories, and *ab initio* theories, and different basis sets have been used in an attempt to achieve an accurate but less computationally demanding method.

In 2006, we showed that CCSD(T) [61–64] is a highly effective method for calculating the change in gas phase Gibbs free energies for deprotonation [13]. CCSD(T) is one of the most effective ways to include electron correlation, which results from the fact that when a particular electron moves, all other electrons tend to move to avoid that particular moving electron. Hartree-Fock theory solves the Schrödinger equation for an average electronic potential; including electron correlation is essential for obtaining meaningful energetic values, and different ways of doing so consume much of the field of computational chemistry. In this case, the coupled cluster calculations included triple excitations for both the complete fourth-order Møller-Plesset (MP4) and CCSD(T) energies (for instance, by using the E4T keyword in Gaussian). The single-point CCSD(T) energy calculations used the augmented correlation consistent polarized n-tuple zeta basis sets (aug-cc-pVnZ, n = D, T, Q, 5) of Dunning [65]. These calculations were performed upon geometries obtained using fourth-order Møller-Plesset perturbation theory [66] with single, double, and quadruple substitutions [MP4(SDQ)]. These optimizations, and their corresponding frequency calculations, employed the aug-cc-pVTZ basis set. The frequency calculations ensured that all structures were optimized to a true energetic minimum on the potential energy surface, and the unscaled thermochemical corrections were used to obtain the zero-point energies, enthalpies, and Gibbs free energies. Furthermore, to estimate the energy at the complete basis set limit, a series of two-point extrapolations on the correlation energy were undertaken [13]. In this scheme (equations 2.23–2.25), an extrapolated value for the correlation contribution to the total energy is obtained using two consecutive correlation energies, $x - 1$ and x, and is then added to a nonextrapolated Hartree-Fock energy [67–69]:

$$E_x^{corr} = E_x^{CCS(ID)} - E_x^{HF} \tag{2.23}$$

$$E_{x-1,x}^{corr} = \frac{x^3 E_x^{corr} - (x-1)^3 E_{x-1}^{corr}}{x^3 - (x-1)^3} \tag{2.24}$$

$$E_{x-1,x} = E_x^{HF} + E_{x-1,x}^{corr} \tag{2.25}$$

The CCSD(T)//MP4(SDQ)/aug-cc-pVTZ method, with the extrapolation to the complete basis set limit using the aug-cc-pVTZ and aug-cc-pVQZ basis sets, yielded a standard deviation of 0.58 kcal/mol when compared to a select set of experimental values of gas-phase deprotonation reactions compiled in the NIST online database, a data set with uncertainty of <1 kcal/mol [13,58]. The low error of the selected NIST data set makes these values extremely useful in determining accurate pK$_a$ calculations and will be referenced throughout this section.

Using model chemistry methods, we also reported ΔG_{gas} calculations with slightly less accuracy, within 1.1 to 1.6 kcal/mol of experimental values. The model chemistries G3 [70], CBS-QB3 [59], CBS-APNO [60], and W1 [71] produced mean absolute deviations of 1.16, 1.43, 1.06, and 0.95 kcal/mol, respectively [10,13]. This result was confirmed in 2005 where G2, G3, and CBS-APNO predicted accurate values of ΔG_{gas} for formation of ion-water clusters when compared to experimental results [72–76]. Contrary to the previous publications, however, CBS-QB3 was less accurate for these clustered ions than the other model chemistries. Although there are many effective methods of gas phase Gibbs free energy calculations, it would be useful to find computationally less demanding methods that produce a similar accuracy [12].

At first glance density functional theory (DFT) appears to offer a more cost efficient approach, although it must be remembered that each particular DFT functional and a given basis set represents its own parameterized method. DFT includes some of the correlation energy, although the exact solution to recover all of it is still unknown and the subject of much theoretical research. For example, fairly accurate results were obtained for PBE1PBE/aug-cc-pVTZ and B3P86/aug-cc-pVTZ [12]. The select NIST data test set included the deprotonation reactions of the following compounds: ammonia, methylamine, dimethylamine, ethylamine, methane, methanol, water, acetylene, ethylene, formaldehyde, hydrogen chloride, propene, nitrous acid, nitric acid, isocyanic acid, furan, and benzene. When compared to experimental results within the NIST database [58], the mean square ΔG_{gas} deviations for PBE1PBE/aug-cc-pVTZ and B3P86/aug-cc-pVTZ were both 1.6 kcal/mol, exhibiting somewhat less accuracy compared to more computationally expensive methods.

In another study, G3MP2, G2, G3, G2MP2, G3B3, G3MP2B3, QCISD(T), CBS-4, CBS-Q, CBS-QB3, and CBS-APNO, produced ΔG_{gas} values for nitrous acid within 0–1.6 kcal/mol [77] of the experimental value of 333.7 kcal/mol [78]. They also found that the less expensive density functional B3LYP produced values within 2.72 kcal/mol of experiment. The commonly used Hartree-Fock level of theory, which does not include correlation energy, produced inaccurate results with a large 4.66 kcal/mol discrepancy [77].

The accuracy of B3LYP has been examined by numerous researchers. In 2003, Fu and colleagues reported that the MP2/6-311++G(d,p) and B3LYP/6-311++ G(2df, p) methods yielded gas phase acidities, or the change in Gibbs free energy of the reaction:

$$AH_{gas} \rightarrow A^-_{gas} + H^+_{gas} \tag{2.26}$$

within 2.2 and 2.3 kcal/mol of experimental values of various organic acids reported in the NIST online database [58,79]. Two years later, Range also reported that B3LYP with the 6-311++G(3df, 2p) basis produced a root mean square error of 2.5 kcal/mol for reaction 2.26, when compared to experimental values from the NIST online database [58]. The article also reported that previously discussed high levels of theory, CBS-QB3, G3B3, G3MP2B3, PBE0, and B1B98, have a root-mean-square error (RMSE) all within 1.3 kcal/mol of experimental values [80]. Reaction 2.26 represents the gas phase dissociation of an acid, which is the top line of Figure 2.1.

Other publications, however, report more accurate values of B3LYP gas phase Gibbs free energy calculations on aliphatic amines, diamines, and aminoamines. In 2007 Bryantsev et al. reported that B3LYP calculations with the basis set 6-31++G** had a mean absolute error of 0.78 kcal/mol from experimental values of the gas phase basicity (ΔG_{gas}) of the reverse reaction of equation 1 reported in the NIST database [58]. This accuracy is comparable to that of expensive, high level model chemistries, but because the experimental values have uncertainties of ±2 kcal/mol, it is difficult to discern exactly how accurate the calculations are in comparison to values in the other publications [81]. The take-home message remains the same: always benchmark DFT calculations for the systems you are interested in computing [52].

2.3   SOLVATION GIBBS FREE ENERGY CALCULATIONS

The largest source of error in pK$_a$ calculations is the change in Gibbs free energy of solvation calculation for the reaction, which is based on the type of solvation model used and the specific level of theory [8,11,15,39,55]. The basic problem is that experimental Gibbs free energies of solvation for ions have error bars of roughly 2–5 kcal/mol, and therefore models that have been developed to reproduce experimental values have the same inherent uncertainty. It is not possible to improve a particular solvation model by simply increasing the basis set, as one can when calculating *ab initio* quantum mechanical gas phase values.

Explicit solvation methods include the addition of solvent molecules directly in the calculation. This method is advantageous because specific solute-solvent interactions are taken into account. These multiple interactions, however, make it more difficult to find a global minimum for the complex [52,55]. The number of necessary solvent molecules included in the reaction also comes into question, leading to the problem of balancing accuracy with computational expense. In fact, although water is one of the most studied molecules in all of science, the actual structure of liquid water is still not settled, and there are no models that accurately replicate all the properties of water [82–84]. It is only quite recently that spectroscopic methods have become advanced enough to determine the structures of small water clusters (n = 6) without recourse to theoretical modeling [85], and the results indicate that high level electronic structure theory does indeed predict the correct ordering of isomers for small water clusters [83–89]. At present highly accurate quantum chemistry methods have been used to calculate accurate structures for clusters of up to 17 water molecules, which is an interesting regime since the innermost water molecule appears to have the properties of bulk water at this point [90–92]. However, the great computational demands of these calculations mean that only a few conformers have been considered, and at this writing it is an open question as to how many explicit waters would need to surround an acid, and then its dissociated products, to enable a purely explicit and accurate water pK$_a$ calculation.

In addition, conformational effects can be daunting; it is difficult to know how many different ion-water configurations are necessary to achieve a conformationally averaged result. Reactions 2.6–2.9 use only one water molecule, but explicit solvation methods can be used to examine the effects of adding additional waters to the reaction:

$$HA_{(aq)} + nH_2O_{(aq)} \rightarrow (H_2O)_n \cdot A^-_{(aq)} + H^+_{(aq)} \qquad (2.27)$$

$$HA_{(aq)} + (n+m)H_2O_{(aq)} \rightarrow (H_2O)_m \cdot A^-_{(aq)} + (H_2O)_n \cdot H^+_{(aq)} \qquad (2.28)$$

$$HA \cdot H_2O(n)_{(aq)} + OH^- \cdot H_2O(m)_{(aq)} \rightarrow A^- \cdot H_2O(n+m)_{(aq)} + H_2O_{(aq)} \quad (2.29)$$

Reactions 2.27–2.29 depict some examples of explicit solvation effects, where n, or n+m, is the number of water molecules used in the reaction. Because of the daunting task of computing enough different configurations with a large number of water molecules, complete with frequency calculations to determine the entropic values necessary to obtain free energies for each configuration, quantum chemistry is not yet used routinely for completely explicit solvent models. Recent evidence, however, suggests that if the standard states for water are included correctly, that the use of a thermodynamic cycle based on reaction 2.28 will yield good values for pK$_a$s if a cluster cycle (and not a monomer cycle) is used for the waters [41].

In contrast to explicit solvation, implicit solvent effects, where actual solvent molecules are not included in the thermodynamic cycle, are easily implemented for pK$_a$ calculations. Various methods used to calculate the change in Gibbs free energy of solvation, such as the Dielectric Polarizable Continuum Model (DPCM) and Conductor-like Polarizable Continuum Model (CPCM), use implicit solvation by constructing a solvation cavity around the molecule of interest. These methods have been shown to compute the Gibbs free energy of solvation for neutral molecules within 1 kcal/mol [52]. The implicit models directly approximate a homogenous dielectric continuum, which represents the response of a bulk solvent. This is computationally less demanding than explicit solvation methods, but it is not particularly accurate for ionic species. Strong electrostatic effects make solvent modeling using implicit solvation more challenging [39]. The method yields less accurate values for these highly charged species [54] and also may impart a false partial positive charge on the system if wave functions penetrate beyond cavity walls [45]. Furthermore, ionic species also have larger Gibbs free energies of solvation, due to solute-solvent interactions. Consequently, a smaller error is required for a charged species to produce the same level of chemical accuracy as a neutral molecule [52]. Aside from problems with ionic species, an additional limitation of implicit solvation models is that the accuracy depends on the selection of proper boundary techniques, such as the type of solvation cavity [54].

While developing an implicit solvation model, Solvation Model 6 (SM6), Kelly, Cramer, and Truhlar found that when calculating the Gibbs free energies of solvation for molecules with concentrated regions of charge densities, more accurate values were obtained by adding explicit waters in addition to the implicit effects of the model. They concluded that this occurred because of significant local solute-solvent interactions, which their implicit model did not take into account [54]. This method of including explicit solvation effects while using an implicit model is referred to as a cluster-continuum model [93] or implicit-explicit model [54], and was pioneered by Pliego and Riveros in 2001–2002 [45,94]. One limitation of this method is that one must determine the number of explicit solvent molecules that yield the most accurate results, which varies based on the type of molecules in the data set [93,95]. The future of pK$_a$ calculations will no doubt include more explicit-implicit approaches, as computers increase in speed and it becomes easier to include multiple conformations for a given system of interest.

Along with deciding whether to use implicit or implicit-explicit solvent models, a specific level of theory and basis set must be used for the calculation of the change in Gibbs free energy of solvation. Similar to the gas phase Gibbs free energy, there are a variety of methods and it can be difficult to determine what combination is the most accurate.

The change in Gibbs free energy of solvation calculation for the reaction is the largest source of error in pK$_a$ calculations. To determine the most accurate method we must look at both the type of solvation model used, implicit, explicit, or cluster-continuum method, and the specific level of theory. As previously mentioned, ionic species, in particular, are extremely difficult to calculate accurately because of their strong electrostatic effects and large Gibbs free energy of solvation values [45,52,54,55].

Implicit solvation models developed for condensed phases represent the solvent by a continuous electric field, and are based on the Poisson equation, which is valid when a surrounding dielectric medium responds linearly to the charge distribution of the solute. The Poisson equation is actually a special case of the Poisson-Boltzmann (PB) equation: PB electrostatics applies when electrolytes are present in solution, while the Poisson equation applies when no ions are present. Solving the Poisson equation for an arbitrary equation requires numerical methods, and many researchers have developed an alternative way to approximate the Poisson equation that can be solved analytically, known as the Generalized Born (GB) approach. The most common implicit models used for small molecules are the Conductor-like Screening Model (COSMO) [96,97], the Dielectric Polarized Continuum Model (DPCM) [98], the Conductor-like modification to the Polarized Continuum Model (CPCM) [99], the Integral Equation Formalism implementation of PCM (IEF-PCM) [100] PB models and the GB SMx models of Cramer and Truhlar [52,57,101,102]. The newest Minnesota solvation models are the SMD (universal Solvation Model based on solute electron Density [57]) and the SMLVE method, which combines the surface and volume polarization for electrostatic interactions model (SVPE) [103–105] with semiempirical terms that account for local electrostatics [106]. Further details on these methods can be found in Chapter 11 of reference 52.

Kelly, Cramer, and Truhlar used the cluster-continuum model in their study of aqueous acid dissociation constants [39]. They compared the correlation between experimental pK$_a$ values and the calculated acid dissociation Gibbs free energies of anions with and without an additional explicit water molecule using SM6. Note that because of the relation between pK$_a$ and ΔG_{aq} as shown in equation 2.1, a plot of pK$_a$ versus ΔG_{aq} should yield a slope of 1/2.303RT, or 1/RTln(10).

The single water molecule was added only to ions containing three or fewer atoms or ones with oxygen atoms bearing a more negative partial atomic charge than that of the water solute. They reported that when only implicit effects were included, a regression equation with a slope of $0.71/RT\ln(10)$ and correlation $r^2 = 0.76$ was computed. When an explicit water was added to the acids, however, the new regression yielded a slope of $0.87/RT\ln(10)$ with a correlation of $r^2 = 0.86$. From this observation, Kelly, Cramer, and Truhlar concluded that, for some anions, the accuracy of acid dissociation energies greatly increased with the addition of one explicit water molecule. Furthermore, they concluded that an implicit model alone cannot produce such accurate results because it does not take into account strong solute-solvent interactions. They argue that previous publications using implicit methods with strong correlations between pK$_a$ values and Gibbs free energy calculations actually have underlying systematic errors in their methods, as indicated by lower slopes [55].

Related to Kelly, Cramer, and Truhlar's conclusion, Klamt, Eckert, and Diedenhofen's 2003 publication studied the correlation between experimental pK$_a$ values and the Gibbs free energies of dissociation for 64 organic and inorganic acids for reaction 2.6. Like the Kelly publication, Klamt, Eckert, and Diedenhofen used an explicit water molecule. Their solvent calculations, however, used Klamt's COSMO-RS level of theory [97]. They reported a correlation of $r^2 = 0.984$ with a standard deviation of only 0.49. The slope of the regression line, however, was fairly low at $.58/RT\ln(10)$. Klamt et al. believed that this discrepancy did not result from the weakness of the calculation method [107]. Another study by Eckert and Klamt, in 2006, confirmed these results by reporting that a correlation of experimental pK$_a$ values with Gibbs free energies of dissociation had an $r^2 = 0.98$ with a deviation of 0.56 pK$_a$ units and again a significantly smaller slope than the accepted $1/RT\ln(10)$ [108].

These values may indicate that COSMO-RS, contrary to Kelly, Cramer, and Truhlar's assertion, is actually more accurate than SM6. However, a more likely possibility is that this lower slope indicates underlying systematic error.

Although Kelly, Cramer, and Truhlar found that a single explicit water molecule increased the accuracy of acid dissociation Gibbs free energies for SM6, this trend was not common to all solvation methods or for the addition of multiple water molecules. They reported the gas-phase binding Gibbs free energies of $(H_2O)_n$ CO_3^{-2} with $n = 0$ to $n = 3$ for SM6, SM5.43R, and DPCM/98 with UAHF (United Atom topological model applied on radii optimized for the HF/6-31G* level of theory) atomic radii levels of theory. As the number of water molecules increased from zero to three, the accuracy also increased for SM6. The two other continuum solvation models, SM5.43R and DPCM, however, decreased in the accuracy of gas-phase binding Gibbs free energies as the number of explicit water molecules increased. With the addition of one water molecule, SM5.43R did become more accurate. It got significantly worse, however, as the number of water molecules increased, and the absolute deviations went from 1 kcal/mol with one water molecule to 10 kcal/mol with three molecules. The most accurate calculation for DPCM was with no explicit water molecules, and the calculations continued to become less accurate with increased numbers of waters surrounding CO_3^{-2}. Overall, the most accurate method for the study was SM5.43R with one explicit water molecule, outperforming SM6 with three water molecules, which had an absolute deviation of 3 kcal/mol [39].

To supplement their previous publication, Kelly, Cramer, and Truhlar also conducted an extensive study surrounding the absolute aqueous solvation Gibbs free energies of ions and ion-water clusters containing a single water molecule. They reported the following mean unsigned errors using values from their recent [55] and previous publication [54]:

Table 2.1 shows that SM6 outperformed all continuum models, with SM6/MPW25/6-31G(d) producing the lowest Mean Unsigned Error (MUE) of 3.3 kcal/mol when used with clustered ions. The data also show that the clustered ions resulted in lower MUE than the unclustered ions for all SM6 calculations by about 1 kcal/mol, reaffirming their conclusion derived from their prior publication [39]. Other levels of theory, however, do not produce as conclusive results and do not always produce lower MUE when implementing the cluster pair approximation. Overall, Kelly, Cramer, and Truhlar concluded that SM6 with diffuse basis functions and clustered ions produce the most reliable values for the absolute aqueous solvation Gibbs free energies ($A^-_{(gas)} \rightarrow A^-_{(aq)}$) [39]. This implies that this method would also lead to the most accurate ΔG_{sol} values. Furthermore, as the parameters in SM6 were originally developed using the $\Delta G_{sol}(H^+)$ value of –264.3 kcal/mol [51], this was a significant finding because it showed that the SM6 calculations are also accurate when combined with the currently accepted $\Delta G_{sol}(H^+)$ value of –265.9 kcal/mol [39,42].

Jia et al. also studied the cluster-continuum method using PCM with the HF/6-31+G(d), HF/6-311++G(d,p), and B3LYP/6-311++G(d,p) levels of theory. For a data set of 5 organic acids, they found that the accuracy of the pK_a calculations increased as the number of explicit water molecules increased from 0 to 3 [109]. In this study, relative pK_a values were computed, so that lack of electron correlation in the gas phase calculation apparently canceled.

Focusing on implicit calculations, da Silva, Kennedy, and Dlugogorski compared the success of DPCM and IEF-PCM in pK_a calculations at the

TABLE 2.1
Mean Unsigned Errors in Absolute Aqueous Solvation Gibbs Free Energies of Ions and Ion-Water Clusters, with a Single Water Molecule, for Various Continuum Solvent Models

Solvent Model	Clustered Data Set[a]	All Ions[b]
SM6/MPW25/MIDI!	3.7	4.8
SM6/MPW25/6-31G(d)	3.3	4.5
SM6/MPW25/6-31+G(d)	3.5	4.6
SM6/MPW25/6-31+G(d,p)	3.5	4.5
SM6/B3LYP/6-31+G(d,p)	3.6	4.7
SM6/B3PW91/6-31+G(d,p)	3.5	4.6
SM5.43R/MPW25/6-31+G(d,p)	5.3	6.1
DPCM/98/HF/6-31G(d)	5.8	5.7
DPCM/03/HF/6-31G(d)	13	14.3
CPCM/98/HF/6-31G(d)	6	6
CPCM/03/HF/6-31G(d)	7.3	7.1
IEF-PCM/03/HF/6-31G(d)	7.4	7.2
IEF-PCM/03/MPW25/6-31+G(d,p)	8.6	8.4

Source: Adapted from references [39] and [54].

[a] Gas-phase optimized geometries at the B97-1/MG3S level of theory.

[b] Gas-phase optimized geometries at the MPW25/MIDI! level of theory.

HF/6-31G(d) level using these polarizable continuum solvent models with UAHF radii and 15 different levels of calculations: HF, MP2, QCISD(T), B3LYP, G1, G2, G2MP2, G3, G3MP2, G3B3, G3MP2B3, CBS-4, CBS-Q, CBS-QB3, and CBS-APNO, for the ΔG_{gas} calculation. The HF, MP2, and QCISD(T) theories use the 6-311++G(3df, 3pd) basis set. Overall, they found that DPCM was more successful than IEF-PCM at calculating the pK$_a$ value of nitrous acid. The most successful method was DPCM with the gas phase calculation at B3LYP/6-311++G(3df, 3pd), which produced an error of only 0.3 pK$_a$ units. Da Silva, Kennedy, and Dlugogorski however, concluded that this was probably because of a cancellation of errors. The other accurate values were calculated using high level theories: CBS-APNO, CBS-QB3, and G2 [77].

Using the DPCM method, da Silva, Kennedy, and Dlugogorski also examined the effect of different basis sets with the HF level of theory using gas phase geometries from G2, CBS-Q, CBS-QB3, and CBS-APNO calculations. They reported that as the basis set size increased, excluding aug-cc-pVTZ, the accuracy of the pK$_a$ calculation also increased. The most accurate basis set paired with HF was aug-cc-pVQZ, which produced an absolute average error of 0.39 pK$_a$ units. Da Silva also studied DPCM with DFT methods instead of HF. The average pK$_a$ values were calculated using G2, CBS-Q, CBS-QB3, and CBS-APNO for ΔG_{gas} values. The Gibbs free energy of solvation was calculated using B3LYP, TPSS, PBE0, B1B95, VSXC, B98, and O3LYP with 6-31G(d), 6-311++G(3df, 3pd), aug-cc-pVDZ, aug-cc-pVTZ, and aug-cc-pVQZ basis sets. Results indicated that the use of DFT methods produces much more accurate results than HF, with all theories within 0.3 pK$_a$ units of experimental values. The most accurate methods were VSXC, TPSS, B98, and B1B95, all with absolute average errors of less than 0.15 pK$_a$ units. Unlike the HF results, da Silva found no benefit in using larger basis sets with DFT [77]. This observation rings true as DFT methods are semiempirical and each method with a given basis set is its own distinct model chemistry suitable for specific systems [110]. Increasing the basis set size does not systematically improve the results as it does in *ab initio* quantum chemistry [12,52].

In 2007, Sadlej-Sosnowska compared the DPCM, CPCM, and IEF-PCM levels of theory for the Gibbs free energy of solvation calculations. The three methods were observed at the HF and B3LYP levels of theory with basis sets 6-31+G*, 6-311++G**, pVDZ, and pVTZ. DPCM was used with a UAHF radius and IEF-PCM was paired with UAHF and UA0. Sadlej-Sosnowska found that IEF-PCM with UAHF was more accurate than DPCM with UAHF when applied to neutral molecules, in contrast to the da Silva, Kennedy, and Dlugogorski results [77]. The most accurate level of theory was IEF-PCM with UAHF at the HF/cc-pVTZ level. In comparing radii, IEF-PCM with UAHF was more accurate than UA0 [111].

Takano and Houk studied several computational methods of solvation calculations and various cavity models in 2005 [44]. They found that for the calculation of aqueous solvation Gibbs free energies the CPCM method at the HF/6-31G(d)//HF/6-31+G(d) and HF/6-31+G(d)//B3LYP/6-31+G(d), with UAKS cavities that have radii optimized with PBE0/6-31G(d), was the most accurate, and produced mean absolute deviations of 2.6 kcal/mol. This mean absolute deviation was based on calculations concerning neutral and charged species. The accuracy of each cavity, UAKS, UAHF(G03), UAHF(G98), BONDI, PAULING, UA0, and UFF, varied based on the type of molecule. CPCM with a UAKS cavity model was also compared to COSMO, SM5.42R, PCM, IPCM [112], and cluster-continuum with MP2/6-31+G(2df, 2p)//HF//6-31+G(d,p) and IPCM. The CPCM data in Takano's publication were compared with SM5.42R, PCM, IPCM, and cluster-continuum data from Pliego and Riveros [94] and COSMO data from various publications [96,113,114]. CPCM was found to have the highest accuracy with a mean absolute deviation from experimental aqueous solvation Gibbs free energies of 3.04 kcal/mol [44]. The other methods had about a 10 kcal/mol deviation, except IPCM with a MAD of 20 kcal/mol [44].

Yu, Liu, and Wang also studied the effect of cavity models. Their pK$_a$ calculations were based upon a system containing an explicit water molecule. Wang and colleagues studied the effect of UAHF, UAKS, Pauling, and Bondi cavity models on the accuracy of CPCM pK$_a$ calculations at the B3LYP/6-311++G(2df, 2p) level of theory on a B3LYP/6-31+G(d) optimized geometry. They reported the pK$_a$ calculation depends greatly on the choice of solvation cavity. The most accurate methods were CPCM with an UAKS or UAHF cavity, which produced mean absolute deviations of 0.38 and 0.40 pK$_a$ units, respectively, from experimental pK$_a$ values [115]. This correlates well with the 2005 data reported by Takano and Houk [44].

Namazian and Halvani studied pK$_a$ calculations with an explicit water using the B3LYP/6-31+G(d,p) level of theory for Gibbs free energy calculations in the gas phase and PCM/B3LYP/6-31+G(d,p) with the UA0 radius for solvation calculations. Using a data set of 66 acids, they found the method accurate within an average of 0.58 pK$_a$ units. The thermodynamic cycles used an explicit water, as in equation 2.6. Although the method produced pK$_a$ values within 0.6 pK$_a$ units, there is some uncertainty in the Gibbs free energy of solvation of H$_3$O$^+$ and the B3LYP level of theory is not the most accurate for Gibbs free energy calculations in the gas phase [116].

Gao and colleagues studied several methods of solvation calculations [117]:

S1: CPCM/HF/6-311+G(d,p)//HF/6-311+G(d,p)
S2: CPCM/B3LYP/6-311+G(d,p)//HF/6-311+G(d,p)
S3: CPCM/HF/6-311+G(d,p)//CPCM/HF/6-311+G(d,p)
S4: CPCM/B3LYP/6-311+G(d,p)//CPCM/HF/6-311+G(d,p)
S5: CPCM/HF/6-31G(d) (Radii-UAHF)//HF/6-311+G(d,p)
S6: CPCM/HF/6-31G(d) (Radii-UAHF)//CPCM/HF/6-311+G(d,p)
S7: SM5.4/AM1 calculated from Spartan 04
S8: SM5.4/PM3 calculated from Spartan 04
S9: SM5.4/AM1 taken from ref. 118 (AMSOL)
S10: SM5.4/PM3 taken from ref. 118 (AMSOL)
S11: SM5.43R/mPW1PW91/6-31+G(d)//mPW1PW91/MIDI!
S12: CPCM/HF/6-31+G(d) (Radii-UAKS)//HF/6-311+G-(d,p)
S13: CPCM/HF/6-31+G(d) (Radii-UAKS)//CPCM/HF/6-311+G(d,p)
S14: Monte Carlo QM/MM

They found that methods CPCM/HF/6-31G(d) (Radii-UAHF)//CPCM/HF/6- 311+G(d,p), SM5.4/PM3 calculated from Spartan 04, SM5.4/AM1 taken from ref. 118 (AMSOL), and Monte Carlo QM/MM produced Gibbs free energy of hydration values within 1 kcal/mol of experimental values. The Gibbs free energy of hydration for the anion, A$^-_{gas}$→A$^-_{aq}$, would be a vertical line on the right side of the thermodynamic cycle shown in Figure 2.1.

The CPCM/HF/6-31G(d) (Radii-UAHF)//CPCM/HF/6-311+G(d,p) and Monte Carlo QM/MM levels of theory were the most accurate and produced Gibbs free energies of hydration within 0.4 kcal/mol of experimental values [117].

Recognizing that the continuum solvent calculations are the weakest link in pK$_a$ calculations, Ho and Coote used the CPCM (with UAKS and UAHF radii), SM6, IPCM, and COSMO-RS models to predict pK$_a$ values for a common data set of neutral organic and inorganic acids [37]. They used four different thermodynamic cycles, and in general found that the COSMO-RS, CPCM, and SM6 models worked best depending on the thermodynamic cycle used.

2.4 PITFALLS AND LESSONS FROM THE LITERATURE

Various thermodynamic cycles can be used in pK$_a$ calculations. Although previously a source of confusion in the field, it is now clear that as long as the most accurate experimental values are used, and no explicit water molecules are added, the choice of cycle should just be a matter of convenience. The most common is based on equation 1.1 and is diagramed in Figure 2.1, where a molecule is simply deprotonated, yielding a corresponding base and the proton in solution [8,9,11]. This cycle depends on the accuracy of the continuum model used to determine the anion (reaction 1.1) or cation (reaction 2.5) solvation energies, calculations that vary in accuracy depending on the system in question and the parameterization of the solvation model.

Other thermodynamic cycles used in various calculations contain explicit water molecules. The effectiveness of this implicit-explicit method in terms of calculating the Gibbs free energy of solvation was discussed in the preceding section. Since the Gibbs free energy of solvation is the largest source of error in determining pK$_a$ values, the accuracy of this calculation often determines the validity of various thermodynamic cycles.

Many thermodynamic cycles contain anions, which often leads to a large error in the computational calculation of the Gibbs free energy of solvation of the anion. As a result, cycles with water molecules or additional acids [38,79,116,117,119,120] are often used to try and remove these sources of error. If accurate Gibbs free energy values are used, pK$_a$ calculations can be fairly accurate, but many papers report pK$_a$ calculations with less accurate Gibbs free energy values for H$^+$. These publications would need to be recalculated with more accepted values to produce reliable and accurate data.

For example, in 2006, Nino and co-workers studied the efficiency of model chemistries G1 [121], G2 [122], and G3 [70] for pK$_a$ calculations [120].

They reported that G1 is the most accurate of the three models, producing an average difference of 0.51 pK$_a$ units from experimental values when used with the CPCM solvation model in the study of aminopyridines. Results showed that the models decreased in accuracy, G1>G2>G3. The publication, however, contains the older accepted $\Delta G_{sol}(H^+)$ value of –264.61 kcal/mol in the 1M standard state, obtained from the acetic acid system [8,11], which as we have discussed in Section 2.11 is 1.29 kcal/mol more positive than the currently accepted –265.9 kcal/mol reported by Tissandier et al. in 1998 [42] and confirmed by Kelly, Cramer, and Truhlar in 2006 [55]. This difference creates a discrepancy in those pK$_a$ values calculated in the paper and the actual values produced by the model chemistries. Changing the value for $\Delta G_{sol}(H^+)$ from –264.61 to –265.9 produces new pK$_a$ values that are approximately 0.95 pK$_a$ units less than those reported by Caballero. This changes the order of accuracy of the methods. After calculating new pK$_a$ values and comparing them to the experimental values reported in the Caballero publication, we find that the previously reported least accurate method, G3, is now the most effective in producing reliable pK$_a$ values [120]. This correction outlines the importance of using correct experimental values and how a difference of merely 1.29 kcal/mol can change the conclusions about the most efficient method of pK$_a$ calculation. G3 has been shown to be superior to G2 (and G2 to G1) for many different gas-phase processes, including deprotonation, so the new ordering makes more sense. Quite simply, Nino and colleagues were led astray by the use of the accepted value for ΔG_{sol} (H$^+$) at the time they started their work.

The significance of the Gibbs free energy of solvation of a proton is also apparent in the publication by Bryantsev, Diallo, and Goddard [81]. In this work the ΔG_{sol} (H$^+$) value was treated as a parameter and fitted in order to obtain the most accurate pK$_a$ values. The Goddard group used $\Delta G_{sol}(H^+) = $ –267.9 kcal/mol and –267.6 kcal/mol for solution-phase and gas-phase optimized calculations, respectively, for the 1 M standard state. The values are off from the accepted value of –265.9 kcal/mol, but still within the 2 kcal/mol error bars assigned by Kelly, Cramer, and Truhlar [55]. Nevertheless, because of this discrepancy, the reported accuracy of less than 0.5 pK$_a$ units for solution-phase optimized geometries and greater than 0.5 pK$_a$ units for calculations on gas-phase optimized geometries might be overstated. These results will change when the accepted $\Delta G_{sol}(H^+)$ value is used.

Examples of excellent work in the literature, where the authors have taken care in their use of the best experimental value for the Gibbs free energy of solvation of the proton, and made careful use of thermodynamic cycles with attention to standard state issues, and made a good choice of methods for gas-phase and solution-phase calculations are listed here. These include the trimethylaminium ion [123], carboxylic acids [124], substituted pyridinium ions and pyridinyl radicals [125], the nitrous acid-ium ion [126], a series of oxicams [127], a series of perfluoroalkyl carboxylic acids [128], monoprotic and diprotic pyridines [129], biologically important carbon acids [14], nucleobases [130,131], a study of neutrals that included methanol, acetic acid, phenol, acetonitrile, methanethiol, toluene, thiophenol, acetone, and ammonia [132], aliphatic amines, diamines, and aminoamides [81], fluorescein and its derivatives [133], nitrous acid [77], and six members of the angiotensin-I-converting enzyme (ACE) inhibitor family [134].

Another pitfall of pK$_a$ calculations concerns unintended consequences when software developers change parameters, those making previous results irreproducible. An example of this concerns the change in the UAHF parameters for the Self-Consistent-Reaction-Field (SCRF) (solvation) calculation in Gaussian03 compared with Gaussian98. In Gaussian98, PCM and CPCM calculations using the UAHF parameters had a particular atom type represented as a carbon bonded to one atom with a double bond, and two other atoms with a single bond, defined as an sp3 hybridized carbon, with a radius of 1.545 Å. In Gaussian03 this atom type was corrected to an sp2 hybridized carbon, with a radius of 1.725 Å. Since Gaussian did not make a big announcement of this change, very few investigators would have discovered on their own that the default values for the SCRF program had been changed, yet the impact on the calculation of Gibbs free energy of solvation values for carboxylic acids and phenols was up to 2 kcal/mol. In another example from Gaussian, significant changes were made in the default SCRF parameters from Gaussian03 to Gaussian09. The impact of these changes, published by Gaussian and explained in the user's reference manual, is highlighted for isopropylamine [135]. Changes like this are the reason why whenever chemists publish results from computational chemistry, they should specify the program version number as specifically as possible, so that other users can validate published results [52]. It is also best practice to provide the results of coordinates, energies, and frequencies in supplemental information.

Recent work on using explicit waters in cluster-continuum or implicit-explicit thermodynamic cycles shows much promise, as long as the standard state issues for water are consistent [37,41]. The key point is that water as a solvent, and water as a solute, and all species involved in the thermodynamic cycle, must be in a 1 M standard state. At this point it is not clear how many explicit waters should be used in a cycle [37], although use of the variational method to determine the number of waters to be used, and putting the waters together as clusters instead of monomers, appears to have much promise [41].

A very interesting report is for a trio of dicarboxylic acids, oxalic ($C_2H_2O_4$), malonic ($C_3H_4O_4$), and adipic ($C_6H_{10}O_4$), all of which can lose two hydrogens on their way to forming a dianion [136]. Continuum models fail dramatically for these three cases, yet four different mixed implicit-explicit models resulted in mean unsigned errors of 0.6–0.8 log units. The authors used –265.9 kcal/mol for the 1M standard state value of $\Delta G_{sol}(H^+)$, the M06-2X [137,138]/MG3S [139] method for the gas-phase calculations, and the SMD [57] and SM8 [102] continuum models for the solvation calculations. They found the best results for the thermodynamic cycle based on reaction 2.30,

$$HA(H_2O)_n(aq) \rightarrow H^+(aq) + A^-(H_2O)_n \qquad (2.30)$$

where the optimal value for n was 4, 3, and 2 waters for oxalic, malonic, and adipic acids, respectively. The explanation is that an isolated carboxylate group only needs one water for improvement in the implicit-explicit scheme, but when two carboxylates are close together, extra microsolvation is required to account for their proximity. In this study, oxalic acid is surrounded by waters, malonic acid only needs a bridge, and for the larger adipic acid one water is enough [136].

Another interesting article using a mixed implicit-explicit approach is a study on protonated amines [140]. A protonated amine reactant is positively charged while the conjugate base product is neutral, so the authors sought the best way to model reactions 2.5 and 2.7. After an exhaustive study, they found that a thermodynamic cycle based on reaction 2.31 gave them results with mean unsigned errors less than 0.6 pK$_a$ units:

$$HA^+(H_2O) + 3H_2O \rightarrow A + H_3O^+(H_2O)_3 \qquad (2.31)$$

The authors used the value of –265.89 kcal/mol for the 1 M standard state value of $\Delta G_s (H^+)$, the M05-2X [138]/6-311++G(d,p) method for the gas-phase calculations, and the IEF-PCM continuum model for the solvation calculations. The same group published a similar study on large phenolic derivatives [141]. Other groups have used the implicit-explicit approach to study the pK$_a$s of formic and carbonic acids [142].

2.5 CONCLUDING REMARKS ON ABSOLUTE pK$_a$ CALCULATIONS

Many variables affect the accuracy of pK$_a$ calculations. In regards to the Gibbs free energy calculation in the gas phase, extra computational expense might be necessary to achieve values within 1.0 kcal/mol for a given deprotonation reaction. However, getting this right is straightforward. CCSD(T) single-point energy calculations, extrapolated to the complete basis set limit, on MP2 or MP4 geometries with decent sized basis sets are accurate within a half kcal/mol or better. DFT methods should be benchmarked against appropriate experimental or *ab initio* results to ensure that the DFT method of choice is suitable for the systems of interest; we recommend the recently developed methods of Zhao and Truhlar [138]. Compound model chemistry methods such as G3, CBS-APNO, and W1 are also highly accurate.

For the Gibbs free energy of solvation calculation, however, it is difficult to discern the most accurate method. Recently, there have been numerous publications exploring the use of the cluster-continuum method (also termed implicit-explicit) with anions. In regard to implicit solvation, there are no definite conclusions on the most accurate method, yet for the PB models the Conductor-like models (COSMO; CPCM) appear to be the most robust over the widest range of circumstances [52]. At this writing the SMx (SMD, SM8, SM12) methods [57,143,144] developed by Cramer and Truhlar seem to be the most versatile, as they can be used by themselves, or with the implicit-explicit model, and the error bars for bare and clustered ions are the smallest of any continuum solvation method. The ability to add explicit water molecules to anions and then use the implicit method (making it an implicit-explicit model) improves the results more often than the other implicit methods that have been published in the literature to date.

Concerning thermodynamic cycles, the most important component is the treatment of the Gibbs free energy of the hydrogen ion. Even a slight difference in values can produce drastically different trends in pK$_a$. The most accurate experimental value should be used in the equation. As of this writing the best values for the experimental Gibbs free energies of solvation, for a standard state of 1M and 298K, are −265.3, −265.6, or −265.9 ± 1 kcal/mol for H$^+$; −104.7 kcal/mol for OH$^-$, −110.3 kcal/mol for H$_3$O$^+$, and −6.32 kcal/mol for H$_2$O [41–43,55]. These values are all consistent with each other, as can be seen by using them in thermodynamic cycles to calculate the dissociation of water into its component ions, where ΔG_{gas} is obtained from the NIST website. Most researchers use the −265.9 value for the proton. For the classic thermodynamic cycle displayed in Figure 2.1, using the −265.9 accepted value for ΔG_{sol}(H$^+$) and considering the conversion of gas-phase calculations to the 1 M standard state (equation 2.22), pK$_a$ values for the reaction in equation 1.1 at 298.15 K can be determined using equation 2.32, with the four calculated energies in kcal/mol:

$$pK_a = [G_{gas}(A^-) - G_{gas}(HA) + \Delta G_{sol}(A^-) - \Delta G_{sol}(HA) - 270.28567]/1.36449$$

$$(2.32)$$

Extra caution should be taken when performing pK$_a$ calculations on highly ionic species, as their strong electrostatic effects and large Gibbs free energies of solvation make accurate calculations difficult. Cycles involving explicit water molecules have their merit when dealing with these compounds. Interested readers should refer to the recent literature to ensure they correct for the standard state of water, which should be 1 M and not 55.34 M in all cycles [37,41]. To further complicate matters, various functional groups or acidic/basic strength of the molecules may also affect the accuracy of methods. If the implicit solvent method used in the calculation of $\Delta G_{sol}(A^-)$ and $\Delta G_{sol}(HA)$ is believed to yield good results for the species in question, then using thermodynamic cycle 1 of Figure 2.1 and equation 2.32 is the most straightforward way to calculate pK$_a$ values. Investigators are encouraged to use the highest level of theory they can afford to calculate $G_{gas}(A^-)$ and $G_{gas}(HA)$.

Finally, it is interesting to note the great success that pK$_a$ calculations combined with experimental results have for elucidating structural phenomena. An excellent example concerns the acid-base properties of tetracyclines. Tetracyclines are of great pharmaceutical interest, and a recent study used pK$_a$ values computed from DFT on the protonated, neutral, and anionic tetracyclines to identify the deprotonation sites and products of each deprotonation [145]. In this study chemical accuracy wasn't necessary, and the combination of B3LYP and PCM calculations was good enough because of the available experimental information. There are other studies in the literature of biologically important molecules, where absolute accuracy in pK$_a$ calculations is not necessary, that have yielded great insight into the protonation state of drugs such as structurally diverse α_1-adrenoceptor ligands [146], and of the flavonoid luteolin and its methylether derivatives [147].

Due to the numerous potential cycles using explicit molecules, levels of theory, basis sets, and types of molecules, it is impossible to determine one specific method that produces the most accurate pK$_a$ values. Rather, this chapter serves to summarize the current literature and illustrate various schemes that have been successful. Accurate attention to detail and the use of benchmark calculations or experimental values to assist in determination of the correct method to use for a particular system are highly recommended. Further research on thermodynamic cycles using explicit cycles, clustered water structures, conformational effects, and advances in continuum solvation calculations will continue to advance this field.

3 Relative pK$_a$ Calculations

The reliance on experimental or calculated values for the Gibbs free energy changes associated with the proton present a significant challenge in predicting accurate pK$_a$ values using the thermodynamic cycles outlined in Chapter 2. However, if the pK$_a$ of a related molecule is accurately known, then the pK$_a$ of a similar molecule can be determined with a relative calculation that eliminates the necessity of using the values for $G_{gas}(H^+)$ and $\Delta G_{sol}(H^+)$ [7,15]. In addition, it is possible to use nuclear magnetic resonance (NMR) to experimentally determine pK$_a$ shifts [148,149]. In relative pK$_a$ calculations, the values for $G_{gas}(H^+)$ and $\Delta G_{sol}(H^+)$ cancel out, so any inaccuracies in their values are irrelevant. Consider two molecules of the same class, HA and HB. We can write a thermodynamic cycle for each of these (Figures 3.1 and 3.2).

Suppose we know the pK$_a$ for HA. Then we can write the pK$_a$ for the unknown HB as follows:

$$pK_a(HB) = pK_a(HA) + \Delta pK_a \tag{3.1}$$

Here ΔpK_a is the difference in pK$_a$ of the two related molecules. Following equation 2.1, $pK_a = \Delta G_{aq}/RT\ln(10)$, so ΔpK_a is given by:

$$\Delta pK_a = \Delta\Delta G_{aq}/RT\ln(10) \tag{3.2}$$

where

$$\Delta\Delta G_{aq} = \Delta G_{aq}(HB) - \Delta G_{aq}(HA) \tag{3.3}$$

Using equations 2.1–2.3 twice, once for HA and once for HB, we have:

$$\Delta G_{aq}(HA) = G_{gas}(H^+) + G_{gas}(A^-) - G_{gas}(HA) + \Delta G_{sol}(H^+) + \Delta G_{sol}(A^-)$$
$$- \Delta G_{sol}(HA) \tag{3.4}$$

and

$$\Delta G_{aq}(HB) = G_{gas}(H^+) + G_{gas}(B^-) - G_{gas}(HB) + \Delta G_{sol}(H^+) + \Delta G_{sol}(B^-)$$
$$- \Delta G_{sol}(HB) \tag{3.5}$$

$$\Delta G_{gas}$$

$$HA_{(g)} \quad \rightarrow \quad A^-_{(g)} \quad + \quad H^+_{(g)}$$

$$\uparrow -\Delta G_{sol}(HA) \quad \downarrow \Delta G_{sol}(A^-) \quad \downarrow \Delta G_{sol}(H^+)$$

$$HA_{(aq)} \quad \rightarrow \quad A^-_{(aq)} \quad + \quad H^+_{(aq)}$$

$$\Delta G_{aq}$$

FIGURE 3.1   Proton-based thermodynamic cycle for molecule HA.

so that equation 3.3 becomes

$$\Delta\Delta G_{aq} = G_{gas}(B^-) - G_{gas}(A^-) - G_{gas}(HB) + G_{gas}(HA) + \Delta G_{sol}(B^-) -$$
$$\Delta G_{sol}(A^-) - \Delta G_{sol}(HB) + \Delta G_{sol}(HA) \tag{3.6}$$

Therefore for relative pK$_a$ calculations, combining equations 3.1, 3.2, and 3.6 the following equation can be used to calculate the pK$_a$ of HB when the pK$_a$ of HA is known:

$$pK_a(HB) = pK_a(HA) + [G_{gas}(B^-) - G_{gas}(A^-) - G_{gas}(HB) + G_{gas}(HA)$$
$$+ \Delta G_{sol}(B^-) - \Delta G_{sol}(A^-) - \Delta G_{sol}(HB) + \Delta G_{sol}(HA)]/RT\ln(10) \tag{3.7}$$

Relative pK$_a$ calculations can be highly accurate, within a few tenths of a kcal/mol, as errors in both the gas-phase and solution-phase values tend to cancel [7]. Numerous highly accurate relative pK$_a$ values have been reported in the literature. For instance, a comprehensive study on a large set of carboxylic acids, alcohols, phenols, and amines using B3LYP for the gas-phase and PCM for the solution-phase calculations spanned over 16 orders of experimental pK$_a$ values, and resulted in standard deviations of 0.37, 0.40, and 0.52 pK$_a$ units for carboxylic acids, alcohols and phenols, and amines, respectively [150]. Nino and co-workers report highly accurate relative pK$_a$ calculations for aminopyridines [120], while Brown and Mora-Diez published moderate relative results for protonated benzimidazoles [38]. A computational study of the acid dissociation of esters

$$\Delta G_{gas}$$

$$HB_{(g)} \quad \rightarrow \quad B^-_{(g)} \quad + \quad H^+_{(g)}$$

$$\uparrow -\Delta G_{sol}(HB) \quad \downarrow \quad \Delta G_{sol}(B^-) \downarrow \Delta G_{sol}(H^+)$$

$$HB_{(aq)} \quad \rightarrow \quad B^-_{(aq)} \quad + H^+_{(aq)}$$

$$\Delta G_{aq}$$

FIGURE 3.2   Proton-based thermodynamic cycle for molecule HB.

and lactones highlights the case study of diketene [151]. Another study reveals that while absolute predictions of four stepwise protonation constants for nitrilotripropanoic acid fail completely, the relative calculations using B3LYP and PCM are accurate within 1 pK_a unit [152]. Borg and Durbeej have determined ground- and excited-state relative pK_a calculations for the phytochromobilin chromophore, revealing the rings that are the strongest acids in the ground and bright first excited singlet states [153,154].

Relative pK_a calculations are the main method used to determine pK_a shifts in biomolecular systems. Such calculations have error bars of several pK_a units for the determination of the pK_a values for amino acid side chains on a protein [155–157]. Because biomolecular systems such as proteins are very large, molecular dynamics simulations use empirical potentials such as AMBER, CHARMM, and OPLS [52]. Correspondingly, solvation models are fitted [158] for use with empirical potentials, such as Sharma and Kaminski's fuzzy-border continuum model [159].

Relative pK_a calculations have also been shown to be highly effective in solvents other than water. Using MP2 single-point calculations on B3LYP geometries for the gas phase, and the PCM continuum model for the solution phase, it was possible to calculate pK_as within 0.5 pK_a units for neutral acids in dimethyl sulfoxide (DMSO), acetonitrile, and tetrahydrofuran solvents [160]. For DMSO, the most polar of these three solvents, addition of one solvent molecule within the PCM formalism (as in Section 2.4) had a significant effect on the accuracy of the results. This is a major finding, as previous high-level absolute pK_a calculations in DMSO solvent without using an implicit-explicit model had error bars of 1.5–2 pK_a units [79,161]. The hydrogen bond network that DMSO makes between reactants and products of proton dissociation reactions affects the pK_a of tertiary alcohols [162].

4 Quantitative Structure-Acidity Methods

We have seen in Chapters 2 and 3 how a "purist" would calculate the pK_a of a chemical compound from first principles, employing a suitable thermodynamic cycle, a reasonably high level of computational theory, and an appropriate solvent model. The attraction of the purist, or absolute approach, is that it relies only on well-established theoretical methods (with the obvious exception of the value for the Gibbs energy of solvation for H^+), and does not depend on the availability of experimental measurements, which may be difficult to perform, missing, or improperly measured. The drawback of the purist, absolute approach is that in order to achieve reasonable accuracy, a high level of theory is normally required, both in the calculation of the compound and in the solvent model. In fact, the absolute approach can be computationally demanding for even a modest level of accuracy, since one must determine the Gibbs energy change for the solution dissociation equilibrium to within ± 1.36 kcal/mol (or ± 5.69 kJ/mol) in order to obtain a pK_a value accurate to ± 1 pK_a unit [8,9,11]. This presents a considerable challenge since although modern quantum chemical techniques can normally determine the aqueous Gibbs energies of solvation for neutral species to within ± 1 kcal/mol, the corresponding accuracies for the accompanying ionic species are typically only about ± 2–4 kcal/mol [37,55]. In addition, the high level of calculation needed for gas-phase calculations, as explained in Section 2.2, rules out absolute methods, at present, for larger molecules. In this chapter we describe an alternative approach that relies on experimental input, but that in turn often supplies reasonable accuracy with less computational demand [36].

4.1 BASIC PRINCIPLES OF THE QSAR APPROACH

Just as detectives seek clues in investigating crimes, so too do scientists seek "clues" in attempting to explain the behaviors of chemicals. The scientific clues generally take the form of molecular features such as partial atomic charges, energy differences, electrostatic potential surface values, and other properties that may be associated with an inclination for a molecule to react, dissociate, or perform some other action of interest. For example, in explaining the preferred sites for electrophilic attack on aromatic frameworks, one typically looks for atomic positions in the aromatic molecules that display negative charges, with the more negative sites presumably being more likely to react in this manner. When the variations in one or more molecular features can be tied numerically to the variations in a particular molecular behavior, the resulting mathematical relation is called a *quantitative structure-property relationship* (QSPR) or, equivalently, a *quantitative structure-activity relationship* (QSAR). Such relationships have found valuable employment in a great variety of applications within physical, biological, and biomedical science [163–170].

For clues regarding the propensity of a molecule to dissociate off a proton, consider the change in pK$_a$ in the series propanoic acid, acrylic acid, and propiolic acid, as illustrated below:

pK$_a$ = 4.87   pK$_a$ = 4.25   pK$_a$ = 1.84

In qualitative terms it is apparent that as the left side of the compound becomes more unsaturated, the compound becomes more acidic (lower pK$_a$). We need, however, to find a quantitative measure that we can tie to the variation in pK$_a$. Since the ionization involves the carboxylic acid group we might first look at the charges on the dissociating hydrogen atom, the dissociating –OH group, the entire carboxylic acid group (–COOH), or the final dissociated carboxylate (–COO$^-$) group, to see if any of these might be closely associated with the pK$_a$ behavior. Of these choices—here using the natural charges calculated using density functional theory—both the charges on the COOH group and those on the OH group appear to follow the pK$_a$ variation reasonably well (see Table 4.1), suggesting the possible use of either Q(COOH) or Q(OH) as a pK$_a$ descriptor. Moreover, it is quite reasonable to suppose that the more loosely the dissociating hydrogen atom is held, as might be indicated by a smaller difference between the charges on the oxygen and hydrogen atoms in the dissociating OH group, the greater will be the tendency for this group to dissociate, giving a lower pK$_a$.

Another possibly useful descriptor might be the energy difference ΔE between the energy of the parent R-COOH compound and its dissociated R-COO$^-$ anionic form. In the present example (Table 4.1) one finds that ΔE also correlates with the tendency of the –COOH group to dissociate. In this case it can be argued that the energy difference ΔE serves as an approximate surrogate for the Gibbs energy difference ΔG between the two species, thus relating directly to the pK$_a$ (see, e.g., references 18, 171, and 172). Other descriptor candidates might also be useful, and the best test would be to assemble a large group of related compounds with their

TABLE 4.1

Some Potential Descriptors for the pK_a of the Propanoic Acid Series

	Q(H)	Q(COOH)	Q(OH)	Q(COO⁻)	ΔE(au)	ΔE(kJ/mol)	pK_a
Propanoic acid	0.513	−0.014	−0.217	−0.84	0.555145	1458	4.87
Acrylic acid	0.519	−0.038	−0.211	−0.844	0.551147	1447	4.25
Propiolic acid	0.52	0.01	−0.178	−0.772	0.536022	1407	1.84

pK_a values and try to find a mathematical relationship between the pK_a values and the candidate descriptors.

The simplest and most straightforward approach is to assume that a linear free energy relationship (LFER) exists between some set of appropriate molecular descriptors [93] and a property of interest P_j—in the present case the pK_a—of the form

$$P_j = pK_a = c_0 + c_1X_1 + c_2X_2 + c_3X_3 + \cdots \qquad (4.1)$$

Here the c_i are the coefficients of the descriptors X_i in the equation. The form (4.1) follows from the basic thermodynamic relationship between the standard Gibbs energy change $\Delta G°$ for a reaction and the reaction's equilibrium constant K_{eq}:

$$\Delta G° = -RT \ln(K_{eq}) \qquad (4.2)$$

In this expression R is the universal gas constant (1.986 kcal/mol-K or 8.314 J/mol-K) and T is the absolute temperature in kelvins. Regarding the estimation of pK_as in particular, the central concept of this approach was emphasized by Perrin, Dempsey, and Sergeant in their seminal 1981 book on pK_a prediction [173]:

> The most useful methods of pK_a prediction are based on the assumption that within particular classes of acids and bases, substituents produce free energy changes which are linearly additive.

In fact, given the limited sophistication of the quantum chemical computations practically accessible at that time, such a QSPR/QSAR approach was the only realistic approach possible.

Relationship 4.1 illustrates the form of a quantitative structure-activity relationship (QSAR) wherein the contributions of different molecular traits, embodied numerically in the descriptors, are added in a linear manner. Frequently just a single molecular descriptor X_1 is used to form the structure-activity relationship, yielding a simple linear regression, and for statistical reasons one generally wants at least five or six data points for each descriptor included in the relationship.

Mathematically equation 4.1 can be solved as a multiple linear regression (MLR) problem, or least-squares fit, and several statistical measures are helpful in evaluating the goodness of the fit between the experimental property values $P_j(Exp)$ and the values provided by the expression on the right-hand side, $P_j(Calc)$. The *correlation coefficient* r shows how closely the two expressions vary with each other, a value of r = 1.0 indicating a perfect fit, r = –1.0 a perfect negative correlation, and r = 0 no correlation. A more revealing measure is r^2, the so-called *coefficient of determination*, which indicates the fraction of the variance in the experimental data that is explained by the mathematical model. Other useful measures include the *standard error* s of the fit, which one wants as small as possible, and the *Fisher statistic* F, for which larger values indicate better fits for a given number of independent variables.

One might also want to assess the "robustness" of the MLR equation, i.e., its freedom from dependence on particular data points that might skew the result. For this one can employ the *covariance* q_{cv}, or its square, q_{cv}^2. Most commonly this value is obtained by a *leave-one-out* (LOO) procedure in which individual data points are sequentially omitted from the regression and new regressions formed, the overall result giving q_{cv}^2, but more elaborate procedures in which several points are simultaneously omitted are also sometimes employed. Visual inspection of a plot of experimental values vs. descriptor values is also helpful in assessing the influence individual data points might assert on the regression.

When an MLR expression is used it is important to determine the standard deviations δc_is of the coefficients in expression 4.1, since these values provide information on the *significance* of the corresponding descriptors X_i. One should therefore express equation 4.1 as

$$pK_a = (c_0 \pm \delta c_0) + (c_1 \pm \delta c_1)\, X_1 + (c_2 \pm \delta c_2)\, X_2 + (c_3 \pm \delta c_3)X_3 + \dots \quad (4.3)$$

The ratio of the coefficient c_i to its uncertainty δc_i is called the *t-test* for that descriptor. Normally one wishes to have the ratio $c_i/\delta c_i$ greater than 3 or 4 to have some confidence that the descriptor is truly contributing to the relationship in a meaningful way. Descriptors with low t-tests should not be included in the regression equation.

When sufficient experimental data values are available, a useful strategy is to divide the data set into a *training* or *model set*, which is used to derive an equation of the form 4.3, and a *test set*, which can be used as an independent check on how well the mathematical model embodied in the model equation performs. One should be careful to ensure that the training set includes compounds with sufficiently broad diversity in the descriptor values to cover the variations in the test set.

In attempting to derive QSPRs/QSARs it is helpful to remember that both models and experimental data have their shortcomings. The statistician George E.P. Box has famously stated, "Essentially, all models are wrong, but some are useful" [174]. Models by their very nature are approximations, or metaphors, for more complex systems [175]. In the case of QSPRs/QSARs the regression model may fall short because the descriptors employed are not adequate or optimal, or simply because the property being modeled is not easily reduced to linear expressions. One should also be aware that not all experimental data reported in the literature are correct. *Outliers*—points falling far from the curve represented by the overall model—may indeed be the result of shortcomings of the model employed, but they may also be due to shortcomings in the experimental measurements or their translation into databases [176,177]. It is one of the virtues of the mathematical models that they can be used not only to predict unmeasured property values, even for compounds not yet produced, and to understand the fundamental features underlying molecular behaviors, but also to alert one to questionable examples in already measured experimental values [178,179].

Cronin, Dearden, and their colleagues have offered advice on how to avoid common pitfalls and mistakes in forming QSARs [180,181].

4.2 HAMMETT AND TAFT CONSTANTS

A great many different types of molecular descriptors have been employed in the formulation of QSPRs and QSARs [170]. In early pK_a studies simple empirical adjustments were employed to represent the influences of structural changes on the pK_a values of reference compounds [182]. A more systematic approach was proposed by Louis P. Hammett, who proposed a linear free energy model for the effects of substituents on the dissociation of benzoic acid [183–185]. Hammett established the general relationship

$$\text{Log}\,(K_{eq}'/K_{eq}°) = \sigma\rho \qquad (4.4)$$

where K_{eq}' is the equilibrium constant for the substituted acid, $K_{eq}°$ is the equilibrium constant for benzoic acid, σ (*sigma*) is the substituent constant for the particular substituent, and ρ (*rho*) is a parameter representing the particular reaction in question. Hammett distinguished between the effects of *meta* (σ_m) and *para* (σ_p) substituents. Applied to pK_as the Hammett equation becomes

$$pK_a = pK_a° - \rho\,\Sigma\sigma_i \qquad (4.5)$$

where the σ_is are the Hammett constants for the substituents present in the substituted acid. This assumes that the substituent effects can be added independently. Later Taft introduced modified substituent constants (ρ^*, σ^*) and terms to account for steric and field effects [186–188].

An early study of cyclic amine acidities by Hall [189] using Hammett constants produced good linear plots for the pK$_a$s, while a study of aliphatic amines using Taft constants also produced linear plots, with separate lines for the primary, secondary, and tertiary amines [190]. Ballinger and Long [191] found a linear correlation between the pK$_a$s of a number of alcohols and Taft σ^* values, with $\rho^* = 1.42$ for the dissociation reaction. Perrin [182] later summarized acidity results using Hammett constants for benzene derivatives in the expressions

Benzoic acids:

$$pK_a = 4.20 - 1.00 \, \Sigma\sigma_i$$

Phenols:

$$pK_a = 9.92 - 2.23 \, \Sigma\sigma_i$$

Anilines:

$$pK_a = 4.58 - 2.90 \, \Sigma\sigma_i$$

At the time it was not common to list statistical values for the regressions. In 1981 Perrin, Dempsey, and Sergeant published their general review of pK$_a$ prediction methods [173], containing an extensive list of substituent constants. A later review by Hansch, Leo, and Taft gave an expanded discussion of Hammett constants along with resonance and field effects that may be associated with these constants [192]. In 1990 Taft and co-workers [193] showed that excellent regressions could be obtained for the gas-phase acidities of a diverse set of 25 acids using the Taft constants.

Overall Hammett and Taft constants often perform well as descriptors for correlating pK$_a$ values and other properties with molecular structure. This has motivated a number of studies, in addition to the Hansch et al. study, that have been designed to uncover the intrinsic molecular features responsible for this success. Early semiempirical theories that have received broad use are based on the Modified Neglect of Differential Overlap (MNDO) method. These models, in order from the first developed to the last by the Dewar and Stewart groups, are MNDO, AM1, and PM3 [52]. An early series of papers by Jaffé [194–196] described a correlation between Hammett constants and pi-electron densities. Gilliom, Beck, and Purcell [197] found a correlation between the Hammett and Taft constants and the reciprocal of the energy of the highest occupied molecular orbital ($1/\varepsilon_{HOMO}$), calculated using the semiempirical

MNDO method. Schüürmann [198] has used PM3 semiempirical calculations to examine the relationship between electronic properties and Hammett constants. Haeberlein et al. [199] found good linear correlations between features of the electrostatic potential surfaces of *para*-substituted anilines and Hammett σ_p constants. Kim and Martin [200] used CoMFA analysis and AM1 partial atomic charges to estimate Hammett σ values for a set of 49 substituted benzoic acids. Using density functional theory (DFT) calculations Takahata and Chong [201] found a correlation between σ values and core-electron binding energy shifts, while Rincón and Almeida [202] examined the possible role of steric effects in these values. Further studies of the connection between Hammett constants and quantum chemical parameters have been carried out by Ertl [203], Sullivan, Jones, and Tanji [204], Gross and Seybold [17], Gross et al. [26], and Simón-Manso [205].

The Hammett constants, since they generally yield good correlations, provide a convenient benchmark for the performance of alternative indices in modeling the pK_as of different sets of compounds. Gross and Seybold showed that several quantum chemical descriptors outperformed the Hammett constants in modeling the pK_as of substituted anilines [17] and phenols [18]. For substituted benzoic acids, however, Hollingsworth, Seybold, and Nadad [20] found that although the quantum chemical indices still provided good correlations with the pK_as, the Hammett constants, which were to a great extent derived from benzoic acid data, were superior ($r^2 = 0.999$).

4.3   THE SEARCH FOR USEFUL QUANTUM CHEMICAL DESCRIPTORS

Indices taken from quantum chemical calculations of electronic structure were recognized early as potentially effective structure-property descriptors, and as theoretical calculations became more sophisticated, this promise became more realizable [167]. Among such descriptors the partial atomic charges on the dissociating groups were seen as important possibilities. However, charge is not an experimental observable, and solving the Schrödinger equation does not yield atomic charges. Since the partial atomic charge on an atom in a molecule is not a proper quantum chemical variable, but an ad hoc construct, a number of different calculation schemes based on different approaches have been proposed to generate atomic charges. In 1993 Wiberg and Rablen [206] examined the utility of six charge schemes for estimating molecular dipole moments from electrostatic potential surfaces. They cautioned that "no one procedure is 'best' for all purposes," but expressed a preference for atoms-in-molecules (AIM) charges [207] when working with dipole moments. (See, however, the criticism of AIM charges by Perrin [208].)

Cramer [52] has divided partial atomic charges into four classes: Class I charges are obtained by means of some scheme, often empirical, other than direct quantum calculation, Class II charges are derived from quantum chemical orbitals, Class III charges are derived by partitioning the wave function or probability distribution, and Class IV charges are obtained by adjusting Class II or III charges to correspond to some observable property.

A number of early investigations employed charge descriptors to establish relationships with acidities. For example, in 1993 Dixon and Jurs [209] used empirically determined atomic charges to estimate the pK$_a$s of a large collection of oxyacids. Citra [210] used partial atomic charges and O-H bond orders determined using the AM1 semiempirical method of Dewar et al. [211] and found good MLR fits ($r^2 = 0.84-0.96$) for sets of phenols, carboxylic acids, and alcohols. Citra used the partial atomic charges on the oxygen and hydrogen atoms as well as the O-H bond order as regression variables. Later Tehan and co-workers [212,213] used the AM1 method to estimate the pK$_a$s for a large collection of compounds, including phenols, carboxylic acids, amines, and heterocyclic compounds. They examined the effectiveness of a number of potential regression descriptors, including partial atomic charges, energy descriptors, and several variables derived from the frontier orbital theory [214], such as the electrophilic superdelocalizability. The latter descriptor, in particular, showed a strong correlation with the pK$_a$s. Jelfs, Ertl, and Selzer [215] also used the electrophilic superdelocalizability, along with partial atomic charges and molecular tree structured fingerprints, to examine the acidities of a number of nitrogen-containing compounds. La Porta and co-workers [216] have used a modified frontier orbital method to examine the acid-base chemistry of organic amines.

Zhang, Kleinöder, and Gasteiger [217] have examined large data sets of carboxylic acids and alcohols using empirical atomic charge descriptors and a multiple linear regression approach. For 1122 aliphatic carboxylic acids they obtained $r^2 = 0.813$ and $s = 0.423$ using five descriptors, and for 288 alcohols they found $r^2 = 0.817$ and $s = 0.755$ using four descriptors. They devised an atomic charge descriptor $Q_{\sigma,0}$ that models the inductive (electron-withdrawing) effect of nearby substituents and displays a strong ($r^2 = 0.848$) correlation with the Taft σ^* constants in the aliphatic carboxylic acids.

Returning to the question of which schemes for partial atomic charges perform best in modeling acidities, Gross, Seybold, and Hadad [24] examined the effectiveness of seven of the most popular charge models—AIM [207], electrostatic potential (ESP) fitting [218], generalized atom-polar tensor (GAPT) [219], Mulliken [220], natural population analysis (NPA) [221], Löwdin [222], and topological Gasteiger charges [223,224]—in modeling the pK$_a$s of substituted anilines and phenols. The calculations were carried out at the density functional theory B3LYP/6-311G** level. In this study, the most effective charges in modeling the pK_as were the NPA, AIM, and Löwdin charges. For the anilines, the charge variations on the aniline nitrogen atom correlated most closely with the pK_a variations. For the phenols, the charges Q(OH) on the phenolic hydroxyl group correlated best with the pK_a variations. More recently these same authors compared the distributions rendered by several popular charge models with the corresponding electrostatic potential surfaces for an extensive set of halogenated hydrocarbons [25]. In general, the different charge models produced quite different values for the partial atomic charges in these compounds, and variations of the electrostatic potential surfaces of the compounds showed that the charge distributions are in fact more nuanced than might be surmised from the partial atomic charge values alone.

Vařeková et al. [225] compared five different partial atomic charge schemes for their effectiveness in modeling the pK_as of substituted phenols. They tried natural (NPA), Mulliken, ESP, Löwdin, and Hirschfeld charge models and several different levels of theory. Overall, their findings were similar to those of Gross et al., with the Mulliken, NPA, and Löwdin charges performing best overall for most levels of theory. They employed a larger data set than earlier studies and also used multiple descriptors in the regressions, thereby achieving higher r^2 values. Vařeková et al. also found that some of the smaller basis sets, such as 6-31G* and STO-3G, with the MP2 and HF approaches, performed well in comparison to larger basis sets.

In addition to charge descriptors, two energy descriptors have been extensively used for QSAR studies. They are the energy difference ΔE (or enthalpy difference ΔH) between the protonated and deprotonated forms of the acid, and ε_{HOMO}, the energy of the highest occupied molecular orbital of the acid. As early as 1979 Catalán and Macias [171] used INDO calculations [226] to establish a relationship between calculated ΔE values and the gas-phase acidities of *meta*- and *para*-substituted phenols. In 1985 La Manna, Tschinke, and Paoloni [227] used STO-3G Hartree-Fock calculations to demonstrate a relationship between ε_{HOMO} and the gas-phase acidities of a series of benzoic acid derivatives. Sotomatsu, Murata, and Fujita [228] used AM1 calculations to establish relationships between both energy and charge descriptors and the gas-phase acidities of substituted benzoic acids. At the same time Grüber and Buss [229] used MNDO and AM1 energy descriptors to establish good MLR correlations with the pK_as of a large set of phenols and carboxylic acids.

In 1993 Adam [93] used density functional theory (DFT) calculations (PW91/6-311+G**) with Bader's AIM charges [207] and the COSMO solvent model [96] and found excellent results for the pK_a values of carboxylic acids, phenols, and other compounds using the ΔE descriptor. Later Gross and Seybold [18], also using DFT (B3LYP/6-311G**) calculations, observed that both ΔE and ε_{HOMO} correlated strongly with the pK_a variations in these compounds. Hollingsworth, Seybold, and Hadad [20] found similar good results for substituted benzoic acids. Studies of aliphatic amines [19], biophenols [21], and azoles [22] have also established good correlations between pK_as and these energy descriptors so long as suitable solvent models were applied.

Zhang, Baker, and Pulay have developed an efficient pK_a estimation protocol based on DFT (OYLP/6-311+G**) calculations of the deprotonation energy ΔE along with the COSMO solvent model [172]. They found that application of a reliable solvent model was crucial to the success of their approach, whereas the particular DFT functional employed and the basis set played more modest roles. These workers applied their procedure to a wide range of compound classes, including carboxylic acids, phosphonic acids, phenols, alcohols, hydroxamic acids, oximes, and thiols, with impressive results [230].

Murray, Politzer, and their co-workers have developed several descriptors based on features of the molecular *electrostatic potential surface* (EPS) that can be used to characterize a variety of chemical and physical properties, including pK_as [26,199,231]. In studies of the acidities of substituted azoles and anilines they showed that values of the most negative surface potentials (V_{min}) and the *minimum local ionization energy* on the molecular surface ($\bar{I}_{S,min}$) showed strong correlations ($r \approx 0.97$) with the pK_as of these compounds. Later, Ma et al. [27] found that $\bar{I}_{S,min}$ and several other EPS descriptors provided good models of the pK_a variations in substituted phenols and benzoic acids. Sakai and co-workers [232] have shown that V_{min} yields an excellent fit ($r^2 = 0.996$) for the aqueous pK_as of a set of 22 amines. These studies demonstrate that features of the molecular electrostatic potential surfaces of acids can offer useful guides for pK_a estimation.

A different tack has been taken by Han, Tao, and their co-workers [233–235]. They used DFT and MP2 (6-311++G**) calculations and first examined bond lengths, vibrational frequencies, and energetic variables of the parent compounds, finding a correlation between the O-H bond lengths of chloro-, bromo-, and fluoro-phenols and their pK_as—longer O-H distances, as occur, e.g., in *ortho*-chlorophenols, being associated with lower pK_as. Recognizing that the acidity of a compound in a solvent depends on the compound's tendency to pass its proton to the solvent, they later explored the idea of employing a hydrogen-bonding "probe" molecule, such as water or ammonia, to estimate the pK_as of various classes of organic compounds [236,237]. They examined structural characteristics of the probe-compound complexes and identified strong correlations between the O-H vibrational frequencies, O-H and C-O bond lengths, and other properties of the complexes and the experimental pK_a values for both phenols and carboxylic acids.

Chattaraj et al. have used a descriptor they term a "group philicity index" ω_g, to estimate pK_as [238,239]. This descriptor is based on concepts developed by Parr, Szentpály, and Liu [240]. Chatteraj and co-workers have applied this index to the estimation of the pK_as of carboxylic acids and phenols [238], finding correlations of r = 0.86 and r = 0.88, respectively. In another study they examined the use of a variety of descriptors in assessing the pK_as of a diverse set of aromatic acids [241].

An interesting, and often unrecognized, feature of QSPR/QSAR methods is that under some circumstances the regression results may be better than the data from which they are constructed. As Gauche [179] has explained, this can occur because the regression line cuts through the (hopefully) random errors in the data to produce a truer dependence of the property on the descriptors.

4.4 ALTERNATIVE APPROACHES

A number of pK_a estimation methods do not fit cleanly in either the first-principles "absolute" category or the QSPR/QSAR category, but warrant attention as alternatives to these approaches. For example, Kim and Martin [200] used the *Comparative Molecular Field Analysis* (CoMFA) of Cramer, Patterson, and Bunce [242], which is based on molecular docking and field estimations, to study the pK_as of imidazoles and related compounds. Later, Gargallo et al. [243] used CoMFA methods to estimate the pK_as of nucleic acid components. Xing and colleagues [244,245] employed "molecular tree fingerprints" along with the partial least squares (PLS) technique [246] to analyze effects of molecular superstructures, and thereby estimate the pK_as of a large number of organic compounds. Lee, Yu, and Crippen [247] have described a decision tree approach to pK_a estimation that displays excellent statistics ($r^2 = 0.94$) for a large data set compiled from the Lange and Beilstein databases.

Milletti and co-workers [248] used descriptors generated by the program GRID [249] to predict the pK_as of large sets of acidic nitrogen compounds and six-membered N-heterocyclic bases, with impressive statistical results. Several groups have employed neural network approaches in estimating pK_as. Luan and co-workers [250] employed neural networks to obtain estimates for the pK_as of 74 neutral and basic drugs. Habibi-Yangjeh et al. have applied several types of neural network analyses to study the pK_as of several classes of compounds, including benzoic acids [251], phenols [251,252], substituted acetic acids [253], and a wide variety of nitrogen-containing compounds [254].

4.5 FREE AND COMMERCIAL PROGRAMS

A number of free and commercial packages for modeling pK_as and other properties are available. Many of these include some dependence on QSPR methods. One example is the SPARC program (Sparc On-Line Calculator, http://ibmlc2.chem.uga.edu/sparc) developed by the Environmental Protection Agency and the University of Georgia. This program employs a reaction center/perturber model with several options, and has achieved good results in estimating the pK$_a$s of 4300 compounds [255] and also a large number of compounds of pharmaceutical interest [255]. Other available packages include ACD/pK$_a$ DB (Advanced Chemistry Development, http://www.acdlabs.com/products), ADME Boxes, ADMET Predictor, Epik, Jaguar, Marvin, Pallas pKalc, and Pipeline Pilot.

Liao and Nicklaus [256] have evaluated the performances of these packages in predicting the pK$_a$s of 197 pharmaceutical compounds, and found the top three performers for these compounds to be ADME Boxes 4.9, ACD/pK$_a$ DB 12.0, and SPARC 4.2, respectively. Yu et al. [257] have compared different methods for calculating pK$_a$s for 1143 organic compounds, including oxygen acids and nitrogen bases. The best performers overall for this large data set were the ACD and SPARC packages. Manchester, Walkup, and You [258] compared ACDLabs/pK$_a$, Marvin (ChemAxon), MoKa (Molecular Discovery), Epik (Schrodinger), and Pipeline Pilot for their ability to estimate the experimental pK$_a$s of 211 drug-like compounds, and found the performances of ACD, Marvin, and MoKa to be essentially equal, while Epik and Pipeline Pilot gave results farther off from the experimental values. Shelley, Calkins, and Sullivan have commented on this analysis and described ways to improve the performance of Epik [259].

Dearden et al. [260,261] have compared different commercial software programs for their ability to predict pK$_a$ and other properties. In their test of 653 compounds ADME Boxes performed best, although VCCLAB and ACD/Labs also performed well. To be safe, they recommend that predictions be obtained from three sources and an average taken when a calculated pK$_a$ is needed.

5 Oxyacids and Related Compounds

In this chapter we describe applications of quantitative structure-activity relationship (QSAR) methods to oxyacids, including alcohols, phenols, carboxylic acids, phosphonic acids, hydroxamic acids, silanols, and thiols. What these acids have in common is that dissociation of the acid proton occurs from an oxygen-hydrogen (or an analogous sulfur-hydrogen) bond. When not otherwise referenced the pK_a values cited have been taken from the *CRC Handbook of Chemistry and Physics* [262].

5.1 ALCOHOLS, PHENOLS, AND CARBOXYLIC ACIDS

Although alcohols, phenols, and carboxylic acids each have special characteristics, we shall consider these compounds together in this section because many published studies have examined these compounds as a group, and it is more efficient to consider these reports in one place.

Aliphatic alcohols typically are weak acids with pK_as around 16, although some branched alcohols, such as 2-butanol ($pK_a = 17.6$ [263]) and *tert*-butanol ($pK_a = 16.9$ [264]–19.2 [263]), have somewhat higher reported values. The presence of electronegative or electron-withdrawing substituents on the alcohols can dramatically lower the pK_a value, as, e.g., in trifluoroethanol ($pK_a = 12.37$), perfluoro-tert-butanol ($pK_a = 5.2$ [265]), and the extreme case of trinitroethanol ($pK_a = 2.37$ [266]). Phenols generally have lower pK_as and greater acidities than aliphatic alcohols. The pK_a of phenol itself is 9.99, and here too electron-withdrawing substituents can significantly lower the pK_a, as, e.g., in 4-nitrophenol ($pK_a = 7.15$) and 2,4,6-trinitrophenol ($pK_a = 0.3$) [267]. Carboxylic acids are, as a rule, considerably more acidic than either alcohols or phenols. The prototype carboxylic acid, acetic acid, has a pK_a of 4.76, and benzoic acid has a pK_a of 4.20.

Some discussion has occurred over the fundamental causes of the above acidity differences. Scientists seek to qualitatively understand phenomena such as acidity, and this has led to a plethora of terms to explain why one particular type of molecule in a class differs from another. The greater acidity of phenols compared to aliphatic alcohols is commonly attributed to delocalization of the negative charge of the phenolic anions over the aromatic framework, in contrast to the more localized charges in the aliphatic anions [193]. Among other things, charge delocalization stabilizes the phenolic anion and reduces the ordering of the aqueous solvent around it compared to the solvent surrounding the more localized charges of the aliphatic anions. This acts to favor the entropic driving force of the dissociation in the phenolic case.

A great deal of attention has also focused on why carboxylic acids are stronger acids than alcohols [268,269]. For example, the gas-phase acidity of formic acid is 37 kcal/mol greater than that of methanol [270], and the aqueous pK_as are 3.75 for formic acid and 15.5 for methanol. The conventional wisdom has been that this results from resonance stabilization of the formic acid anion. In the late 1980s Siggel and Thomas [271–273] challenged this view, arguing that the enhanced acidity of carboxylic acids was due mainly to normal inductive (electron-withdrawing) effects. Some [269,270] have presented studies supporting the viewpoint of Siggel and Thomas, whereas others [268] have opposed this viewpoint. Studies by Rablen [269] and Holt and Karty [270] appear to indicate that roughly two-thirds to three-quarters of the enhanced acidity of carboxylic acids can be attributed to inductive or electronic stabilization and only a minor portion to resonance. Wiberg, Ochterski, and Streitwieser [274] have explained the greater acidities of carboxylic acids compared to enols in terms of stabilizing effects involving charge transfer to the carbonyl group.

Calder and Barton [275] have called attention to the important role of solvation in driving the thermodynamics of carboxylic acid dissociation. They note that the enthalpy term $\Delta H°$ for the dissociation of carboxylic acids in water is generally small compared to the $-T\Delta S°$ term, so that the large negative entropy change—mostly due to solute-solvent interactions—dominates these dissociations. Adam [93] has also noted this feature, which contrasts with the enthalpy-driven dissociations characteristic of amines.

One of the earliest systematic efforts to find correlates between the acidities of alcohols and direct structural properties was the observation by Goulden of a linear relation between the aqueous pK_as of phenols and their OH vibrational frequencies measured in carbon tetrachloride solution [276]. For carboxylic acids Goulden found three separate lines for the pK_a-v(OH) plots, depending on the nature of the residues connected to the COOH group.

A number of early studies also found good linear relationships between the pK_as of alcohols and the sum of their Hammett or Taft substituent constants, σ or σ^* [191,277,278]. Takahashi et al. found that the pK_as of a large collection of alcohols could be estimated using either Taft substituent constants or the carbonyl frequencies of their esters [265]. Taft and co-workers [193] found a strong relationship ($r^2 = 0.999$) between the gas-phase pK_as of a collection of OH acids and empirical substituent constants representing polarizability, field-inductive, and resonance effects. These workers also showed that the same type of analysis could be extended to the acidities of these compounds in dimethyl sulfoxide (DMSO) and water.

Grüber and Buss [229] used a variety of charge and energy descriptors from MNDO and AM1 [211] calculations to derive single and multiple linear regression (MLR) expressions for the pK_as of 190 phenols and carboxylic acids. Best correlations were achieved for the phenols, although satisfactory results were also found for the carboxylic acids.

As early as 1979, Catalán and Macias demonstrated linear relationships between the gas-phase acidities of monosubstituted phenols and benzoic acids and a variety of molecular indices calculated by Intermediate Neglect of Differential Overlap (INDO) and *ab initio* methods [171]. Among these indices the calculated energy difference $\Delta E = E(AH) - E(A^-)$ provided especially good linear relationships with the acidities. La Manna, Tschinke, and Paoloni reported a linear relationship between the gas-phase acidities of substituted benzoic acids and the highest occupied molecular orbital (HOMO) energies of the corresponding anions [227].

In 1993 Dixon and Jurs [209] used empirically calculated atomic charges to describe the pK_as of a diverse collection of alcohols, phenols, and carboxylic acids. They found a strong linear relationship ($r = 0.993$) between their charge parameters and the pK_as for a set of 135 oxyacids. Citra later used charges and bond orders from AM1 calculations to characterize the pK_as of phenols, carboxylic acids, and alcohols [210]. Here too the best correlations were found for the phenols. Citra also compared his MLR results with results from two commercially available programs and found that all three approaches performed well. He cautioned, however, that his molecular orbital results were sensitive to the final optimized geometry found in the calculations.

Tehan et al. employed AM1 calculations [211] to calculate descriptors from frontier electron theory and applied these to estimating the pK_as of phenols, benzoic acids, and aliphatic acids [212]. Among the descriptors examined the electrophilic delocalizability generally provided the best models for the pK_as of these compounds.

Gelb and Alper [279] have investigated the acidities of several haloacetic acids using potentiometric and conductometric methods, and found differences in the results of the two experimental methods, which they attributed to the presence of an ion-associated (H^+A^-) species contributing to the electrolytic conductance. This proposal is consistent with the model of Schwartz [280], which pictures the dissociation of moderately strong acids as a two-step process involving an intermediate ion-associated pair ($H^+ \cdot A^-$):

$$HA \rightarrow (H^+ \cdot A^-) \rightarrow H^+ + A^-$$

Chaudry and Popelier [281] used a method combining "quantum topological molecular similarity descriptors" obtained from Bader's atoms-in-molecules (AIM) [207] approach to analyze the pK$_a$s of carboxylic acids, anilines, and phenols. They obtained good results, with $r^2 = 0.920$ for the carboxylic acids, $r^2 = 0.974$ for the anilines, and $r^2 = 0.952$ for the phenols.

Simons et al. have examined the abilities of several semiempirical methods to estimate the pK$_a$s of substituted phenols, anilines, and benzoic acids [23]. For the phenols the best correlation was found using the descriptor ΔE (kJ/mol) obtained from calculations based on the AM1 method [211]:

$$pK_a = -0.023 \ (\pm 0.002) \ \Delta E + 11.7 \ (\pm 0.2) \quad \text{(phenols)} \quad (5.1)$$

$$n = 19, \quad r^2 = 0.872, \quad s = 0.31, \quad F = 116$$

The best result for the benzoic acids using the RM1 method [282] again used ΔE:

$$pK_a = -0.016 \ (\pm 0.003) \ \Delta E + 5.78 \ (\pm 0.32) \quad \text{(benzoic acids)} \quad (5.2)$$

$$n = 17, \quad r^2 = 0.893, \quad s = 0.11, \quad F = 126$$

Zhang, Kleinöder, and Gasteiger developed multiple linear regression models for the pK$_a$s of large data sets of both aliphatic alcohols and carboxylic acids using newly developed descriptors [217]. The regression variables accounted for inductive effects, accessibility and polarizability of the acidic oxygen, and π-electronegativity of the α-carbon. A five-descriptor model for 1122 aliphatic carboxylic acids yielded $r^2 = 0.813$ and $s = 0.423$, and a four-descriptor model for 288 alcohols gave $r^2 = 0.817$ and $s = 0.755$. It should be noted that aliphatic acids are typically more challenging to model than their aromatic analogs, so that these results should be considered in that context.

Gross, Seybold, and co-workers have applied several different QSAR schemes in examining the pK$_a$s of different classes of oxyacids [18,20,21,24,27]. Among the descriptors found to give good correlations with the pK$_a$s were the energy of the highest occupied molecular orbital, ε_{HOMO}, the proton transfer energy ΔE, and several charge descriptors. For substituted phenols the following relationship was found [18]:

$$pK_a = 49.5 \ (\pm 2.7) \ \varepsilon_{HOMO} + 13.8 \ (\pm 0.2) \quad (5.3)$$

$$n = 19, \quad r^2 = 0.951, \quad s = 0.187, \quad F = 328$$

The energy descriptor ΔE (in kcal/mol, relative to ΔE for phenol) also was strongly associated with the pK$_a$s for these compounds:

$$pK_a = 0.098 \ (\pm 0.06) \ \Delta E + 9.95 \ (\pm 0.006) \quad (5.4)$$

$$n = 19, \quad r^2 = 0.944, \quad s = 0.199, \quad F = 287$$

Electron-donating substituents were seen to yield higher pK_as and electron-withdrawing substituents lower pK_as. In these studies it was found that correction for the zero-point energies did not significantly affect the results, an assumption also suggested in other studies [93]. In addition, the above results were vacuum results, found without any correction for solvent. This is presumably because the solute-solvent interactions remain reasonably uniform within the series of closely related aromatic compounds.

As noted earlier, the assignment of charges to individual atoms in a molecule is not a uniquely defined quantum chemical procedure, and the assignment of such charges rests on somewhat arbitrarily devised schemes. Gross, Seybold, and Hadad [24] examined the abilities of seven popular charge assignment schemes for their ability to correlate with the pK_a variations in substituted anilines and phenols. Among the charge schemes tested, Bader's AIM charges [207], Löwdin charges [222], and natural population analysis charges [221] performed best. Charge descriptors, such as the charges on the dissociating OH group, Q(OH), the entire COOH group, Q(N), and $Q(NH_2)$ generally performed well in modeling the pK_as.

Bioactive phenols found in vegetables, fruits, wines, and other sources are believed to carry health benefits to consumers. Kreye and Seybold modeled the pK_as of a diverse collection of biophenols using both semiempirical RM1 calculations [282] and DFT calculations [21]. They found that both charge and energy descriptors gave moderately good regressions. Of the charge descriptors Q(OH) determined at the B3LYP/6-31+G* level in an SM8 model solvent [102] performed best:

$$pK_a = -8.18 \ (\pm 1.14) - 88.92 \ (\pm 9.12)Q(OH) \qquad (5.5)$$

$$n = 15, \quad r^2 = 0.880, \quad s = 0.97, \quad F = 95$$

The best energy descriptor was $\Delta E_{aq}(SM8)$, the energy difference in the SM8 solvent:

$$pK_a = -96.3 \ (\pm 11.4) + 229.5 \ (\pm 26.4) \ \Delta E_{aq}(SM8) \qquad (5.6)$$

$$n = 15, \quad r^2 = 0.924, \quad s = 1.1, \quad F = 76$$

Chipman [283] has discussed the use of dielectric continuum models for the computation of pK_a values, with particular reference to alcohols and carboxylic acids in water, dimethyl sulfoxide, and acetonitrile. He concluded that whereas continuum solvent models properly account for long-range solvent effects, they are less successful at treating short-range specific interactions.

Adam [93] has employed DFT (PW91/6-311+G**) calculations, Bader's AIM energy E_H (= $E_{HA} - E_{A^-}$) of the dissociating proton [207], and the COSMO solvent model [96,99] to examine the aqueous pK_as of a wide variety of carboxylic acids, phenols, and anilines. Adopting some reasonable assumptions about the equilibrium in solution he obtained the expression

$$pK_a = \Delta G°/2.303RT \approx -E_H/2.303RT + C, \qquad (5.7)$$

where E_H is in kJ/mol, R is the gas constant, T is the absolute temperature, and C is a constant at a given temperature. At T = 298.15 K the slope 1/2.303RT in expression 5.7 equals −0.1752 mol/kJ when energies are expressed in kJ/mol. Using E_H as a descriptor Adam found generally excellent correlations, ranging from $r^2 = 0.849$ to $r^2 = 0.991$, between the pK$_a$s calculated using expression 5.7 and the experimental pK$_a$s for the different series of compounds studied. He found that inclusion of one or two explicit, hydrogen-bonded water molecules typically improved the regression results, bringing the slopes in expression 5.7 to values close to the theoretical value. For example, for aliphatic carboxylic acids with two explicit waters he found

$$pK_a = -0.1770 \, E_H - 154.671 \qquad (5.8)$$

$$n = 19, \quad r^2 = 0.957, \quad MUE = 0.17$$

where MUE is the *mean unsigned error.* For the carboxylic acids the added waters were found to form characteristic hydrogen-bonded rings, including the carboxylic acid group. For the larger mixed set of aliphatic carboxylic acid dihydrates, substituted benzoic acid dihydrates and substituted phenol monohydrates he found expression 5.9:

$$pK_a = -0.1725 \, E_H - 150.486 \qquad (5.9)$$

$$n = 19, \quad r^2 = 0.991, \quad MUE = 0.217.$$

Ma et al. [27] used features of the electrostatic potential surface (EPS) to model the pK$_a$s of a collection of substituted phenols. They found that the minimum local ionization energy $I_{S,min}$ on the EPS showed a good correlation with the pK$_a$s for these compounds:

$$pK_a = -1.413 \, (\pm 0.124) \, I_{S,min} + 20.73 \, (\pm 0.99) \qquad (5.10)$$

$$n = 19, \quad r^2 = 0.885, \quad s = 0.292, \quad F = 132$$

Han, Tao, and their co-workers [233,234] studied the pK$_a$s of several sets of halogenated phenols and found correlations between the O-H and C-O bond lengths in the phenols, as determined by B3LYP/6-311++G** or MP2/6-311++G** calculations, and the halophenol pK$_a$s. In later studies they examined complexes of water and ammonia with both halophenols and carboxylic acids and found good correlations between the pK_as of the compounds and several properties of the complexes [235–237].

Chattaraj and co-workers [284] have used a "group philicity index" ω_g^+ to estimate the pK_as of a variety of alcohols, phenols, and carboxylic acids. This index derives from earlier work of Parr, Szentpály, and Liu [240]. Using the group philicity index for the COOH group in a set of carboxylic acids, for example, they obtained the relation

$$pK_a = -4.51 \ \omega_g^+ + 7.4 \ \text{(carboxylic acids)} \qquad (5.11)$$

$$n = 31, \quad r^2 = 0.74, \quad s = 0.12$$

Similar relationships were derived for substituted phenols and alcohols. They have also applied this method with additional descriptors to the estimation of the pK_as of some aromatic acids [241].

Harding, Popelier, and co-workers [285,286] have employed a variety of quantum chemical approaches in their estimation of the pK_as of oxyacids. In a study of 228 carboxylic acids they used what they call "quantum chemical topology" to find pK_a estimates. They tested several different methods, including partial least squares (PLS), support vector machines (SVMs), and radial basis function neural networks (RBFNNs) with Hartree-Fock and density functional calculations, concluding that the SVM models with HF/6-31G* calculations were most efficient [285]. For a data set of 171 phenols they found that the C-O bond length provided an effective descriptor for pK_a estimation [286].

It is apparent from several of the above studies that the proton-transfer energy difference $\Delta E = (E_{A^-} - E_{HA})$ is a generally useful descriptor for the estimation of pK_as. Based on this, Zhang, Baker, and Pulay [172] have developed an efficient protocol for pK_a estimation using the DFT OLYP functional [287], a 6-311+G** basis set, and the COSMO solvent model [96,99]. In this protocol geometries are first optimized within the COSMO solvent using a 3-21G* basis set. After testing their protocol [172] they applied the method to a broad range of organic acid classes, including alcohols, phenols, carboxylic acids, phosphonic acids, hydroxamic acids, and thiols [230]. Their models for the present classes included the following:

$$pK_a = 0.3333 \ \Delta E - 90.4470 \ \text{(alcohols)} \qquad (5.12)$$

$$n = 28, \quad r^2 = 0.97, \quad MAD = 0.40$$

$$pK_a = 0.3312 \ \Delta E - 89.7135 \ \text{(phenols)} \qquad (5.13)$$

$$n = 67, \quad r^2 = 0.96, \quad MAD = 0.26$$

$$pK_a = 0.2428 \ \Delta E - 66.6235 \quad \text{(monocarboxylic acids)} \qquad (5.14)$$

$$n = 82, \quad r^2 = 0.93, \quad MAD = 0.20$$

In these expressions the energy differences ΔE are in kcal/mol and MAD is the *mean absolute deviation* from the experimental values. These results represent a high standard for pK$_a$ QSAR applications for oxyacids.

In a follow-up study Zhang [288] has addressed the question of whether the results above can be improved by adding explicit waters of solvation to the continuum solvent model. A special focus was on whether the theoretical value of the regression slope—0.7330 mol/kcal if energies are expressed in kcal/mol (see, e.g., the above study by Adam [93] and also the reports by Klamt et al. [107], Kelly, Cramer, and Truhlar [39], and Eckert, Diedenhofen, and Klamt [289])—could be approached by addition of explicit waters, as earlier studies suggested. Overall Zhang found that although adding explicit waters did increase the values of the regression slopes for different compound classes to values closer to the theoretical value, this result came at the expense of poorer quality of the fits and higher MAD values. From this standpoint the additional computational effort associated with adding explicit waters appears not to be justified. This finding is in sharp contrast to adding explicit waters for absolute pK$_a$ calculations, as outlined in Chapter 2.

Burger, Liu, and Ayers [290] have developed a method based on reference molecules and proton-transfer transition states for the estimation of pK$_a$ values for amines, alcohols, and carboxylic acids. They report a three-descriptor MLR model for 48 oxyacids with $r^2 = 0.993$ and $s = 0.26$.

Recently Vařeková et al. [225] have demonstrated the power of partial atomic charge descriptors in an extensive study of the pK$_a$s of a set of 124 phenols. They examined the performance of seven levels of theory (MP2, Hartree-Fock, B3LYP, BLYP, BP86, AM1, and PM3), with three basis sets (6-31G*, 6-311G, and STO-3G) and five population analysis schemes (natural population analysis [NPA], Mulliken, Löwdin, Hirschfeld, and Merz-Singh-Kollman electrostatic potential). To make the analyses as efficient as possible they performed calculations using 3D structures determined by the CORINA 2.6 software package [291], and then employed three-term MLR fits using the charges on the hydroxyl hydrogen Q(H), the hydroxyl oxygen Q(O), and the α-carbon atoms Q(C1) as descriptors. Twenty-two of their 83 models had $r^2 \geq 0.95$, with standard errors in the range 0.4 to 0.5, and 47 of the 83 had $r^2 \geq 0.90$. All of the theory levels tested performed adequately, with the MP2, HF, B3LYP, and BP86 methods performing best. The Mulliken, Löwdin, and NPA charges showed the closest correlations with the pK$_a$s, while the 6-31G* basis set was the best among the three basis sets tested.

5.2 PHOSPHONIC ACIDS

Phosphonic acids are organic compounds with the general forms shown below, where R, R^1, R^2, and R^3 are normally alkyl or aromatic groups, which may in addition have amino or other functional groups attached

[292,293]. Phosphonic acids tend to be good chelating agents for metal ions, and in natural settings are often found bound to calcium or magnesium ions. A variety of phosphonic acids have been found in plants, single-celled organisms, fungi, mollusks, and other organisms, and some have found use as antibacterial agents [294]. Consequently, their biochemical and environmental activities are of interest [295].

Several theoretical studies have been directed at phosphonic acids, among the earliest being that of Jaffé, Freedman, and Doak [296], who measured the pK_as of 25 *meta-* and *para-*substituted benzenephosphonic acids and utilized Hammett's equation to rationalize the values observed. They subsequently measured the pK_as of 16 *ortho-*substituted compounds and concluded that hydrogen bonds were formed between the phosphono group ($-PO_3H_2$) and many of the *ortho* substituents. The phosphono group could act as either a donor or acceptor.

Guthrie [297] has estimated the pK_as of the O-H and P-H bonds in phosphonic acids and phosphonic esters based on available literature data. For $P(OH)_3$, for example, the pK_a values are estimated as $pK_{a1} = 7.40$, $pK_{a2} = 11.9$, and $pK_{a3} = 16.4$. (The corresponding values for phosphoric acid, H_3PO_4, are $pK_{a1} = 2.16$, $pK_{a2} = 7.21$, and $pK_{a3} = 12.32$ [262].)

Ohta [298] has examined the acidities of a number of alkylphosphonic acids and found a linear relationship between the experimental pK_as and the (Mulliken) electron densities at the phosphorous atom. Ohta examined results from the MNDO, AM1, and PM3 methods, and found the best results ($r = 0.918$) were those determined using the semiempirical PM3 molecular orbital method [299,300]. When halogen-containing substituents were excluded this improved considerably to $r = 0.983$ ($r^2 = 0.966$).

Parthaserathi et al. [238] have used the group philicity index ω_g^+ to estimate the pK_as of several phosphoric acid esters and related compounds. They included the calculations in a group with anilines and alcohols, but the numerical results for the phosphoric acids appear generally good. Moser, Range, and York [301] have calculated gas-phase proton affinities and basicities for a large collection of acyclic and cyclic phosphates.

Zhang, Baker, and Pulay [230] have more recently examined the pK_as of 90 phosphonic and phosphoric acids using their general protocol [172] based on the OLYP/6-311+G** DFT method and the COSMO force field. This led to an excellent fit of the data:

$$pK_a = 0.2451 \; \Delta E - 67.5525 \quad \text{(phosphonic acids)} \qquad (5.15)$$

$$n = 90, \quad r^2 = 0.98, \quad MAD = 0.28$$

Considering the inherent uncertainties in the experimental values, this result provides a strong method for estimation of phosphonic acid acidities.

5.3 HYDROXAMIC ACIDS AND OXIMES

Hydroxamic acids are a class of compounds that contain an N-hydroxyl group bound to an amide group. The general structure is shown below:

They can act as chelating agents and play a number of roles, such as growth factors and iron transporters, in living cells [302,303].

In 1972 Agrawal and Shukla [304] measured the dissociation constants of a number of *para*-substituted benzohydroxamic acids (BHAs) in aqueous solution at 25°C and showed that the pK$_a$s were linearly related to the pK$_a$s of the corresponding *para*-substituted benzoic acids through the relation

$$pK_a \, (\text{BHAs}) = 1.02 \; pK_a (\text{benzoic acids}) + 4.62$$

Also, they found that the BHA pK$_a$s formed a straight line when plotted against the Hammett constants for the substituents.

Zhang, Baker, and Pulay [230] have applied their pK$_a$ estimation protocol [172] to the acidities of hydroxamic acids. For their complete data set of 45 hydroxamic acids they found anomalous results:

$$pK_a = 0.0582 \; \Delta E - 8.7682 \quad \text{(all hydroxamic acids)} \qquad (5.16)$$

$$n = 45, \quad r^2 = 0.08, \quad MAD = 0.50$$

When they separated their set into (–NHOH) and (–NROH) acids, the results improved somewhat, but were still poor compared to their results for other classes:

$$pK_a = 0.2029 \; \Delta E - 51.5091 \quad \text{(–NHOH hydroxamic acids)} \quad (5.17)$$

$$n = 33, \quad r^2 = 0.61, \quad MAD = 0.29$$

$$pK_a = 0.1223 \; \Delta E - 28.6533 \quad \text{(–NROH hydroxamic acids)} \quad (5.18)$$

$$n = 12, \quad r^2 = 0.60, \quad MAD = 0.17$$

For a set of 29 oximes they found good results:

$$pK_a = 0.2280 \, \Delta E - 58.7143 \text{ (oximes)} \qquad (5.19)$$

$$n = 29, \quad r^2 = 0.95, \quad MAD = 0.39$$

5.4 SILANOLS

Silanols are silicon analogs of alcohols, and can be loosely defined as "compounds containing Si-OH bonds" [305,306]. Reviews of these compounds have been given by Lickess [307] and Chandrasekhar, Ramamoorthy, and Nagendran [305]. The latter workers classify silanols into three main groups according to whether the compounds contain Si-OH, $Si(OH)_2$, or $Si(OH)_3$ groups. Silanols have long been of interest because of their importance in chromatography, and have more recently been recognized as catalysts [308,309] and even potential protease inhibitors [305].

Most of the interest in the acidities of silanols has stemmed from their involvement in chromatography [310–313]. In an early study West and Baney [314] examined the acidities and basicities of silanols and concluded that "the silanols studied are much more strongly acidic, but only slightly less basic, than the carbinols with analogous structure." The overall order of acidities of silanols and carbinols decreases in the order aryl silanols > alkyl silanols > aryl carbinols > alkyl carbinols [305]. However, in part because of the considerable variation in surface preparation techniques, the pK_a values reported for silanol groups have covered a wide range [315,316]. Using −OH frequency shifts Hair and Hertl [310] found a pK_a value of 7.1 for a silica surface. In a second harmonic surface study Ong, Zhao, and Eisenthal [317] found two types of silanol sites, with pK_a values of 4.5 and 8.5. Méndez et al. [312] also found two types of surface silanols on liquid chromatography columns, with pK_as of ca. 4.0 and 6.5. Others have likewise found two types of sites [318]. For comparison, the aqueous solution dissociation constants for silicic acid, $Si(OH)_4$, at 30°C are 9.9, 11.8, 12, and 12 [262].

Only a few attempts have been made to use computational methods for the silanol acidities. Sverjensky and Sahai have developed a method for estimating surface protonation equilibrium constants from the surface dielectric constant and an average Pauling bond strength. Rustad and co-workers [319] have used molecular dynamics methods to estimate the pK_a for the reaction $>SiOH \rightarrow >SiO^- + H^+$ at 8.5. Tossell [320] used several quantum chemical levels of theory to establish correlations between calculated gas-phase ΔE and ΔH values and experimental aqueous solution pK_as. Liu et al. [308] have employed both experimental and computation methods to study the gas-phase properties of organic silanols.

TABLE 5.1

Comparison of the pK$_a$s of Some Representative Oxyacids and Thiols

R-OH	pK$_a$	R-SH	pK$_a$
Methanol	15.5	Methanethiol	10.33
Ethanol	15.5	Ethanethiol	9.72
Phenol	9.99	Benzenethiol	6.62
Formic acid	3.75	Thioformic acid	—
Acetic acid	4.756	Thioacetic acid	3.33
Benzoic acid	4.204	Thiobenzoic acid	(3.61)[a]

[a] Kreevoy, Ref. [323]. With permission.

5.5 THIOLS

Thiols are best known for their biochemical activities, especially in protein chemistry, and for their generally unpleasant odors. In addition, the antioxidant glutathione, an assortment of flavoring molecules, and some medicines carry the characteristic mercapto (–SH) group. Thiols are more acidic than their analogous alcohols: some representative pK$_a$s are shown in Table 5.1. It is apparent that the thiol pK$_a$s are typically 3–5 pK$_a$ units lower than their analogous oxycompounds. These differences can be explained at least in part by the lower electronegativity of sulfur (2.58) compared to that of oxygen (3.44) and the lower bond dissociation energy of the S-H bond (353.6 kJ/mol) compared to that of an O-H bond (430 kJ/mol) [321,322].

Apparently the earliest attempt to find a QSAR model for the thiol pK$_a$s was a study by Kreevoy and co-workers [323], who found a linear relationship between the pK$_a$s of 11 thiols and their Taft σ* constants. Two compounds, H$_2$S and thiophenol, were ouliers. The former deviation they explained as resulting from steric effects in solvation, and the latter because of resonance effects.

Bordwell et al. [321] have developed a relationship between the bond dissociation energies (BDEs), oxidation potentials, and the pK$_a$s of weak acids. Chandra, Nam, and Nguyen [324] have used density functional theory (B3LYP/6-311++G**) to determine the BDEs and proton affinities (PAs) of a number of substituted thiophenols and their anions. Greater PA values are expected to correspond to lower acidities.

Zhang, Baker, and Pulay [230] have included thiols in their study of pK_a values using ΔE as a single descriptor. For a data set of 42 thiols they found

$$pK_a = 0.3769\ \Delta E - 103.3833\ \text{(thiols)} \qquad (5.20)$$

$$n = 42, \quad r^2 = 0.95, \quad MAD = 0.34$$

Hunter and Seybold [325] have recently examined additional methods and descriptors for this class of compounds. They found that although the SPARC program generally underestimated the thiol pK_as by about 0.4 pK_a units, it nonetheless provided good correlations for these compounds:

$$pK_a = 0.27\ (\pm 0.30) + 1.02\ (\pm 0.03)\ pK_a(\text{SPARC})\ \text{(thiols)} \qquad (5.21)$$

$$n = 45, \quad r^2 = 0.957, \quad s = 0.40, \quad F = 968$$

The aromatic thiols were very well treated by SPARC ($r^2 = 0.982$) and aliphatic thiols less well ($r^2 = 0.907$). Semiempirical RM1 calculations [282] also performed well using the descriptor $Q(S^-)$, the partial atomic charge on the anionic sulfur atom:

$$pK_a = -6.7\ (\pm 0.6) - 17.8\ (\pm 0.9)\ Q(S^-)\ \text{(RM1 thiols)} \qquad (5.22)$$

$$n = 46, \quad r^2 = 0.938, \quad s = 0.5, \quad F = 665$$

When B3LYP/6-31+G* calculations were used with the SM8 aqueous solvent model [102], a modest regression correlation was found when three outliers were removed:

$$pK_a = 11.9\ (\pm 0.2) - 28.7\ (\pm 1.2)\ Q(SH)\ \text{(DFT-SM8, all thiols)} \qquad (5.23)$$

$$n = 44, \quad r^2 = 0.929, \quad s = 0.5, \quad F = 547$$

6 Nitrogen Acids

Nitrogen acids, including amines and heterocyclic compounds, play many key roles in industry, biochemistry, and pharmacology, making accurate determination of their pK_as an important goal from both experimental and theoretical viewpoints. The common feature of the compounds to be considered in this chapter is that the acid proton dissociates from an N-H bond.

6.1 ALIPHATIC AMINES

As a class, the aliphatic amines encompass a large number of drugs, neurotransmitters, hormones, and other bioactive compounds [326]. In contrast to the oxyacids discussed in Chapter 5, the first dissociation in aliphatic and aromatic amines occurs from a charged species, an ammonium ($-NH_3^+$) group:

$$R\text{-}NH_3^+ \rightarrow R\text{-}NH_2 + H^+$$

It is notable that in the gas phase the basicities of the methyl amines increase in a regular order with methyl substitution, as might be expected from simple considerations of induction and electron donation by the methyl groups, but in aqueous solution the basicities and acidities display an irregular pattern [327,328]. The aqueous acidities fall in the following order:

NH_3	$(CH_3)_3N$	$(CH_3)NH_2$	$(CH_3)_2NH$
$pK_a = 9.25$	9.80	10.66	10.73

These values presumably result from the differing solvations of the methylammonium ions, where even small entropy differences can have a significant effect [329–336].

Bissot, Parry, and Campbell [337] have shown that analogous pK_a variations are found in the methylhydroxylamine and the methylmethoxyamine series. Note that introduction of the hydroxyl group onto ammonia causes the pK_a of hydroxylamine to fall to 5.94. The pK_a of methoxyamine is still lower, at 4.60 [337].

In contrast to the carboxylic acids discussed in Chapter 5, where the dissociation process was largely controlled by entropic factors [275], enthalpic factors dominate the dissociations of the aliphatic amines [93,331]. Recently Hamborg and Versteeg [338] measured the temperature dependences of the dissociation constants and associated thermodynamic properties of a number of amines and alkanolamines.

Morgenthaler et al. [326] have discussed in some detail the dependences of amine basicities on substitution, which allows one to "tune" the amine basicity for lead optimization and other pharmaceutical purposes. In particular, fluorine substitution is especially effective in lowering the amine pK$_a$ due to the strong inductive (electron-withdrawing) influence of the fluorine atoms.

The 1957 study by Hall [190] was the first systematic study of the acid/base behavior of the aliphatic amines. He found that the pK$_a$ values for primary, secondary, and tertiary amines followed roughly parallel straight lines when plotted against Taft's σ^* constants. The trend lines were:

$$pK_a = -3.14 \, \Sigma\sigma^* + 13.23 \text{ (primary amines)} \qquad (6.1a)$$

$$pK_a = -3.23 \, \Sigma\sigma^* + 12.13 \text{ (secondary amines)} \qquad (6.1b)$$

$$pK_a = -3.30 \, \Sigma\sigma^* + 9.61 \text{ (tertiary amines)} \qquad (6.1c)$$

Hall also studied the base strengths of substituted piperidines in water and acetonitrile [189,339]. He found that field effects and steric effects on solvation were important in these compounds.

Several workers have employed features of the molecular electrostatic potential surface to study the pK$_a$s of amines. Nagy, Novak, and Aszasz [340] found good linear correlations between the pK$_a$s of pyridines, anilines, and aliphatic amines and the minima of the electrostatic potentials for these compounds. They used the CNDO/2 method [341] as well as a bond-increment method with data from Perrin's 1972 compilation [342] for their study. Using the bond-increment approach they found the following relationships:

$$pK_a = -0.037 \, V_{min} - 14.45 \quad \text{(pyridines)} \qquad (6.2a)$$
$$n = 8, \quad r^2 = 0.949, \quad s = 0.46, \quad F = 111$$

$$pK_a = -0.085 \, V_{min} - 35.06 \quad \text{(anilines)} \qquad (6.2b)$$
$$n = 13, \quad r^2 = 0.817, \quad s = 0.91, \, F = 41$$

$$pK_a = -0.069 \, V_{min} - 24.40 \quad \text{(aliphatic amines)} \qquad (6.2c)$$
$$n = 20, \quad r^2 = 0.906, \quad s = 0.61, \quad F = 176$$

Nagy [333] also investigated the role of hydration in stabilizing the protonated forms of the amines.

In 2002 Tehan et al. [213] used semiempirical AM1 [211] computations and descriptors from the frontier electron theory [214] to analyze the pK_as of nitrogen acids. For a set of 77 aliphatic amines they found

$$pK_a = -8.01 \text{ SE}_1 - 65.78 \quad \text{(aliphatic amines)} \tag{6.3a}$$

$$n = 77, \quad r^2 = 0.72, \quad s = 0.65, \quad F = 194$$

where SE_1 is the electrophilic delocalizability of the nitrogen atom. Adding a second descriptor, FN_1, the nucleophilic frontier electron density on the nitrogen atom, improved the fit:

$$pK_a = -7.04 \text{ SE}_1 + 3.01 \text{ FN}_1 - 56.97 \quad \text{(aliphatic amines)} \tag{6.3b}$$

$$n = 77, \quad r^2 = 0.81, \quad s = 0.54, \quad F = 158$$

These results, although quite acceptable, do serve to reflect the general difficulty in modeling the pK_as of aliphatic compounds as compared to aromatic compounds.

In 2008 Seybold [19] used both semiempirical and density functional theory (DFT) quantum chemical descriptors to examine the pK_a variations in aliphatic amines. Here the semiempirical RM1 method [282] performed better in modeling the aliphatic amine pK_as than did the more time-consuming DFT B3LYP/6-31G* method. It was found that the vacuum calculations on a data set of 28 aliphatic amines yielded only very modest correlations ($r^2 = 0.712$), but application of the SM5.4 aqueous solvent model [118] and elimination of two unrepresentative compounds greatly improved the RM1 results:

$$pK_a = 42.64 \ (\pm 1.62) - 0.1024 \ (\pm 0.0049) \ \Delta E_{aq} \quad \text{(aliphatic amines)} \tag{6.4}$$

$$n = 26, \quad r^2 = 0.948, \quad s = 0.68, \quad F = 438$$

In this expression the energies are in kJ/mol. The B3LYP/6-31G* calculations on these compounds yielded a reasonable, but less impressive, fit with use of the SM5.4 solvent model:

$$pK_a = -84.1 \ (\pm 5.8) + 198.9 \ (\pm 12.4) \ \Delta E_{aq} \quad \text{(aliphatic amines)} \tag{6.5}$$

$$n = 26, \quad r^2 = 0.915, \quad s = 0.69, \quad F = 257$$

Here the energies are in hartrees.

Zhang, Baker, and Pulay also examined aliphatic amines in their application of their pK_a estimation protocol [172]. They used OLYP/3-21G*/COSMO geometries and then employed the OLYP/6-311+G** calculations with the COSMO solvent model to determine energy differences $\Delta E = (E_{neutral} - E_{cation})$. For a data set of 79 aliphatic amines they found

$$pK_a = -128.4367 + 0.4726\ \Delta E \quad \text{(aliphatic amines)} \tag{6.6}$$

$$n = 79, \quad r^2 = 0.950, \quad SD = 0.50, \quad MAD = 0.35$$

where SD is the standard deviation and MAD is the mean absolute deviation.

6.2 ANILINES

Many more studies have been focused on the pK_as of substituted anilines than on those of the aliphatic amines. In an early study Biggs and Robinson [277] measured the pK_as of a number of substituted anilines and phenols in water and examined the data using Hammett constants as descriptors. For the anilines they found

$$pK_a = 4.580 + 2.889\ \sigma \quad \text{(anilines)} \tag{6.7}$$

$$n = 12, \quad MAD = 0.04$$

In 1992 Haeberlein et al. [199] showed that two descriptors obtained from the molecular electrostatic potential surface, the minimum ionization energy $I_{S,min}$ and the minimum of the molecular electrostatic potential $V_{S,min}$, correlated with the pK_as of *para*-substituted anilines. These descriptors also showed strong correlations with the Hammett constants. More recently Sakai et al. have shown that V_{min} is an exceptionally good descriptor ($r^2 = 0.996$) for the aqueous pK_as of a set of 22 amines. Murray, Brinck, and Politzer [344] showed that these same results could be extended to other nitrogen acids, in particular to nitrogen heterocyclics [231].

Hilal et al. [345] applied the SPARC computational program to estimate the first and second pK_as of 214 azo dyes and related aromatic amines (358 pK_a values in all). They were able to obtain pK_a estimates with roughly the same uncertainty as that for the experimental values.

Koppel et al. [346] compared the acidities of 73 NH acids in gas-phase, dimethyl sulfoxide, and water. Overall they found that the acidities in the different phases were strongly correlated, and that substituent effects on the acidities of these acids were significantly reduced in both solvent phases.

Gross and co-workers examined the pK_as of substituted anilines from several perspectives. In an initial study Gross and Seybold [17] examined a number of structural and other effects of substitution on the aniline framework using B3LYP/6-311G** calculations. A strong correlation was found between the natural population analysis [221] charge $Q_{NPA}(N)$ on the amino nitrogen atom and the pK_a:

$$pK_a = -543\ (\pm 27)\ Q_{NPA}(N) - 437\ (\pm 22)\quad \text{(anilines)}\qquad (6.8)$$

$$n = 19,\quad r^2 = 0.960,\quad s = 0.26,\quad F = 404$$

In a 2002 study Gross, Seybold, and Hadad [24] examined the effectiveness of several charge measures in modeling the pK_as of anilines and phenols. The AIM [207], Löwdin [222], and NPA [221] charges were found to be the most effective measures in this regard. (More recently, Gross, Seybold, and Hadad [25] have compared the charge variations in halogenated hydrocarbons as portrayed by several charge measures, and compared these single-value pictures of the atomic charges with the more nuanced descriptions presented by the electrostatic potential surfaces of these compounds.) Gross et al. also examined the effectiveness of Hammett constants, charges, and descriptors from the electrostatic potential surface (EPS) in correlating the pK_as of 36 anilines. Two descriptors from the EPS, the minimum ionization energy $I_{S,min}$ [199,344] and the minimum of the molecular electrostatic potential $V_{S,min}$ [199], were the best performers:

$$pK_a = -3.02\ (\pm 0.13)\ I_{S,min} + 44.9\ (\pm 1.7)\quad \text{(anilines)}\qquad (6.9)$$

$$n = 36,\quad r^2 = 0.949,\quad s = 0.29,\quad F = 633$$

$$pK_a = -0.154\ (\pm 0.006)\ V_{min} + 7.37\ (\pm 0.46)\quad \text{(anilines)}\qquad (6.10)$$

$$n = 36,\quad r^2 = 0.945,\quad s = 0.30,\quad F = 585$$

Tehan et al. [213] used the AM1 [211] semiempirical method and parameters from frontier electron theory [214] to model the pK_as of 55 anilines. Using the electrophilic superdelocalizability SE_1 of the nitrogen atom they obtained the relation

$$pK_a = -7.83\ SE_1 - 66.64\quad \text{(anilines)}\qquad (6.11)$$

$$n = 55,\quad r^2 = 0.77,\quad s = 0.96,\quad F = 179$$

Milletti et al. [248] have used the GRID software program [249] to predict the pK_as of a large number of nitrogen acids. This approach uses 3D molecular interaction fields (MIFs) from the GRID program as descriptors. A training set of 24,617 pK_a values was used to develop the MIFs for a large number of fragments, and these precomputed values were used taking into account their topological distance from the ionizable site of interest. A partial least squares (PLS) [347] procedure was employed for

33 prediction models pertaining to different ionizable sites. The authors report results of $r^2 = 0.97$ for a set of 421 acidic nitrogen compounds and $r^2 = 0.93$ for a set of 947 six-membered heteroaromatic ring compounds.

Simons et al. have explored the ability of several semiempirical methods to estimate the pK$_a$s of substituted phenols, anilines, and benzoic acids [23]. For the anilines the best correlation was found using single-point DFT/B3LYP/6-311G** calculations based on AM1 [211] geometries:

$$pK_a = -79.5 \ (\pm 3.4) \ Q(NH_2) + 1.15 \ (\pm 0.13) \quad \text{(anilines)} \qquad (6.12)$$

$$n = 19, \quad r^2 = 0.969, \quad s = 0.23, \quad F = 533$$

In this case the best AM1 results used ΔE as a descriptor ($r^2 = 0.886$).

Zhang [343] applied the DFT-COSMO protocol of Zhang, Baker, and Pulay [172] to a set of 37 anilines, with the following result:

$$pK_a = -106.8916 + 0.3961 \ \Delta E \quad \text{(anilines)} \qquad (6.13)$$

$$n = 55, \quad r^2 = 0.981, \quad SD = 0.31, \quad MAD = 0.23$$

Thus it is apparent that a number of QSAR schemes using a variety of descriptors are capable of finding good models for the pK$_a$s of substituted anilines.

6.3   AZOLES AND SOME OTHER HETEROCYCLICS

Heterocyclic compounds are ubiquitous in biology, medicine, and biochemistry, and the azoles form an important class of nitrogen heterocyclic compounds with five-membered aromatic rings that can contain one to five nitrogen atoms (other heteroatoms can also be present). Many drugs and pharmaceuticals are derived from azole structures. The structures of representative azoles are shown in Figure 6.1. Note that some of these compounds exist in tautomeric equilibrium in solution.

A great deal of early work on the acidities of the azoles was performed by Catalán, Abboud, and Elguero [348]. They found, for instance, that annelation increases the gas-phase acidity by 5.9–8.2 kcal/mol [349], and carried out early DFT (B3LYP/6-31G**) calculations of the gas-phase acidities [350]. Lopez et al. [351] found that the pK$_a$ values of pyrazoles and benzazoles display two straight lines when plotted against the oxidation potentials of pentacyanoferrate(II) complexes of these compounds, allowing estimates for missing pK$_a$ values.

In 1991 Brinck et al. [231] demonstrated a relationship between the pK$_a$s of azines and azoles and their molecular surface ionization energies, in the form of the descriptor $I_{S,min}$. For a set of a dozen compounds they found a correlation coefficient of 0.99 between the aqueous pK$_a$ values and values of $I_{S,min}$ for the ring nitrogens.

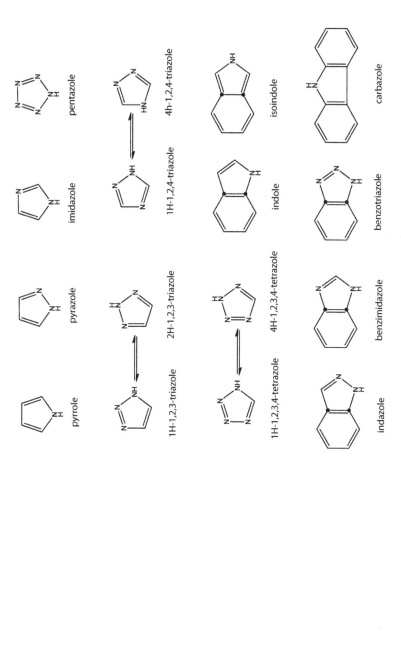

FIGURE 6.1 Structures of some representative azoles (not all tautomeric forms are shown). (Reprinted with permission from P.G. Seybold and W.C. Kreye, *Int. J. Quantum Chem.* **2012,** *112,* 3769.)

Topol, Tawa, and Burt [352] estimated the absolute and relative acidities of substituted imidazoles using DFT and high-level *ab initio* methods with a self-consistent reaction field solvent model. They were able to determine the acidities within an average absolute deviation of 0.8 pK$_a$ units.

In 1998 Koppel et al. [346] compared the acidities of a large collection of NH acids, including some heterocyclic compounds in gas phase, dimethyl sulfoxide, and water. As noted earlier, they found the acidities in the different phases to be strongly related.

A more recent study by Seybold and Kreye [22] of the acidities of alcohols and azoles in these three phases also found strong correlations between the acidities of the compounds in gas phase, DMSO, and water. These workers also performed quantum chemical QSAR calculations using the semiempirical RM1 [282] and DFT B3LYP/g-31+G** methods. The RM1 method performed well in modeling the azoles in the condensed phases, the best descriptor being the energy of the highest occupied molecular orbital (HOMO) of the anion form, $\varepsilon_{HOMO(-)}$:

$$pK_a(DMSO) = 5.98 \ (\pm 0.32) \ \varepsilon_{HOMO(-)} + 37.9 \ (\pm 1.3) \quad \text{(RM1, azoles)} \quad (6.14)$$

$$n = 14, \quad r^2 = 0.967, \quad s = 1.1, \quad F = 348$$

$$pK_a(H_2O) = 5.45 \ (\pm 0.39) \ \varepsilon_{HOMO(-)} + 31.9 \ (\pm 1.5) \quad \text{(RM1, azoles)} \quad (6.15)$$

$$n = 12, \quad r^2 = 0.952, \quad s = 1.25, \quad F = 198$$

The RM1 calculations did not do well in calculating the gas-phase acidities. In contrast, the B3LYP/6-31+G** calculations gave an excellent fit of the gas-phase acidities, showing that the relatively modest level of theory was adequate for this purpose:

$$\Delta G° = 536 \ (\pm 21.3) \ \Delta E + 36.7 \ (\pm 7.3) \quad \text{(DFT, azoles)} \quad (6.16)$$

$$n = 10, \quad r^2 = 0.995, \quad s = 0.51, \quad F = 1701$$

The DFT calculations, when combined with the SM8 solvent model of Marenich et al. [102], also gave good accounts of the acidities in DMSO and water:

$$pK_a(DMSO) = 338 \ (\pm 20) \ \Delta E_{DMSO} - 140 \ (\pm 9) \quad \text{(DFT-SM8, azoles)} \quad (6.17)$$

$$n = 11, \quad r^2 = 0.968, \quad s = 1.16, \quad F = 272$$

$$pK_a(H_2O) = 311 \ (\pm 15) \ \Delta E_{H2O} - 131 \ (\pm 7) \quad \text{(DFT-SM8, azoles)} \quad (6.18)$$

$$n = 12, \quad r^2 = 0.977, \quad s = 0.86, \quad F = 426$$

Using single-point calculations based on the vacuum geometries and the SM8 solvent model gave almost the same results, so that in this case geometry equilibration within the solvent model did not noticeably improve the results.

Zhang [343] applied the previously described pK_a estimation protocol of Zhang, Baker, and Pulay [172] to several sets of heterocyclic bases with impressive results. For a large set of 115 unsaturated five-membered ring bases Zhang found

$$pK_a = -154.0060 + 0.5566\ \Delta E \quad \text{(unsaturated five-membered bases) (6.19)}$$

$$n = 115, \quad r^2 = 0.972, \quad SD = 0.74, \quad MAD = 0.58$$

For a set of 329 five- and six-membered unsaturated ring bases Zhang found

$$pK_a = -147.2277 + 0.5320\ \Delta E \quad \text{(all unsaturated bases)} \quad (6.20)$$

$$n = 115, \quad r^2 = 0.964, \quad SD = 0.67, \quad MAD = 0.52$$

Recall that this protocol uses OLYP/3-21G* geometries combined with single-point OLYP/6-311+G** calculations with the COSMO solvent model. The excellent results obtained appear to support the use of the small basis set for the geometry optimizations, as well as the use of the COSMO solvent model for the aqueous solvent.

In addition to the azoles some pK_a studies have looked at the substituted pyridines. Chen and MacKerell [353] have applied *ab initio* and AM1 calculations [211] to both the gas-phase and aqueous acidities of substituted pyridines. An MP2/6-31G* approach with the IPCM solvent model gave the best results, although the AM1 calculations yielded similar results. Gift, Stewart, and Bokashanga [354] have developed a laboratory experiment using nuclear magnetic resonance (NMR) to measure the pK_as of pyridine and its methyl derivatives.

6.4   AMINO ACIDS

Amino acids are the building blocks of proteins. Because of this connection and the central role they play in biochemistry there has been considerable interest in the physicochemical properties of amino acids, as well as how their properties relate to their biochemical actions. Sjöström and Wold [355], for example, have proposed a relationship between properties of the amino acids and their roles in the genetic code.

It is well known that under normal conditions (pH ~ 7) in aqueous solution and in their crystals amino acids assume a zwitterion form as shown below [356]. Here R is the amino acid side chain that defines the amino acid. Bouchoux [356] has reviewed the gas-phase properties of

$$
\begin{array}{cc}
\text{COOH} & \text{COO}^- \\
| & | \\
\text{H}_2\text{N} - \text{C} - \text{H} \qquad & \qquad ^+\text{H}_3\text{N} - \text{C} - \text{H} \\
| & | \\
\text{R} & \text{R} \\
\text{Neutral form} & \text{Zwitterion form}
\end{array}
$$

amino acids, including gas-phase basicities and proton affinities and also thermodynamic quantities. In the gas phase the isolated amino acids favor the neutral form [356,357]. In fact, in the gas phase the neutral form of glycine, for example, is favored by roughly 70 kJ/mol over its zwitterion form. This makes an absolute calculation (Chapter 2) of the pK$_a$ using thermodynamic cycles impossible, unless one adds enough explicit waters to soak up the zwitterionic charge. Intramolecular hydrogen bonding in the gas phase can also play an important role in stabilizing certain conformers. Gas-phase basicities and proton affinities of amino acids have been compiled by Harrison [358]. Other studies have examined gas-phase properties from both theoretical and experimental viewpoints [301,359,360].

A number of theoretical studies have investigated just how the solvent environment stabilizes the zwitterion (Z) form. Interaction with a single water does not stabilize the Z form of glycine [361], but some MP2 calculations suggest that two waters can stabilize this form [362,363]. Bachrach [364], on the other hand, has used DFT calculations (PBE1PBE/6-311+G**) to examine the sequential solvation of glycine and concluded that seven solvating water molecules are needed for the two forms to become isoenergetic. Calculations with other amino acids suggest varying numbers of waters are required for Z form stabilization for these cases [365,366]. Complexation with metal cations can also stabilize the Z form, as can complexation with ammonium ion [367]. Complexation with I$^-$ can stabilize the Z form of arginine [357].

Compilations of the aqueous pK$_a$s of the carboxylic acid and amino groups of amino acids can be found in several places [262,368] and are given in Table 6.1. The pK$_a$ values for the α-carboxylic acid group range from 1.70 (histidine) to 2.38 (tryptophan) and for the α-amino group from 8.73 (asparagine) to 10.47 (proline). These values compare with the pK$_a$s of acetic acid (CH$_3$-COOH, 4.76) and methylamine (CH$_3$-NH$_2$, 10.66). The relatively small variations found in the carboxyl and amino group pK$_a$s of the amino acids represent the effects of the different local electrostatic environments for each amino acid in solution. Of the 20 standard amino acids commonly found in proteins two (aspartic acid and glutamic acid)

TABLE 6.1
pK$_a$ Values for the 20 Amino Acids (L-Forms) Commonly Found in Nature

Amino Acid	Abbrevation	pK$_{a1}$ (HBC&P)[a]	pK$_{a2}$ (HBC&P)[a]	pK$_{a1}$ (Parrill)[b]	pK$_{a2}$ (Parrill)[b]
Alanine	Ala	2.33	9.71	2.35	9.87
Arginine	Arg	2.03	9.00	2.01	9.04
Asparagine	Asn	2.16	8.73	2.02	8.80
Aspartic acid	Asp	1.95	9.66	2.10	9.82
Cysteine	Cys	1.91	10.28	2.05	10.25
Glutamine	Gln	2.18	9.00	2.17	9.13
Glutamic acid	Glu	2.16	9.58	2.10	9.47
Glycine	Gly	2.34	9.58	2.35	9.78
Histidine	His	1.70	9.09	1.77	9.18
Isoleucine	Ile	2.26	9.60	2.32	9.76
Leucine	Leu	2.32	9.58	2.33	9.74
Lysine	Lys	2.15	9.16	2.18	8.95
Methionine	Met	2.16	9.08	2.28	9.21
Phenylalanine	Phe	2.18	9.09	2.58	9.24
Proline	Pro	1.95	10.47	2.00	10.60
Serine	Ser	2.13	9.05	2.21	9.15
Threonine	Thr	2.20	8.96	2.09	9.10
Tryptophan	Trp	2.38	9.34	2.38	9.39
Tyrosine	Tyr	2.24	9.04	2.20	9.11
Valine	Val	2.27	9.52	2.29	9.72

Note: pK$_{a1}$ refers to the –COOH group and pK$_{a2}$ refers to the –NH$_3^+$ group.

[a] Lide, D. R., *CRC Handbook of Chemistry and Physics*, Vol. 90, Boca Raton, FL: CRC Press, 2009–2010. With permission.

[b] Parrill, A., *Amino Acid pK$_a$ Values* [368]. With permission.

have acidic side chains, and three (histidine, lysine, and arginine) have basic side chains. These are shown in Table 6.2.

Fazary, Mohamed, and Lebedeva [369] have examined the protonation equilibria of the 20 standard α-amino acids in water and dioxane-water mixtures. In a separate study Fazary, Ibrahium, and Ju [370] examined the dependence of the dissociation of L-norvaline, a branched analog of valine, on solvent properties such as ionic strength and temperature, and in several organic solvents.

So far as the authors are aware, no systematic QSAR study of the amino acid pK$_a$s has been carried out, possibly because of the relatively small variations found for the amino and carboxylic acid pK$_a$s observed in these compounds.

TABLE 6.2

pK$_a$ Values for the Amino Acid Side Chains

Amino Acid	Side Chain	pK$_a^a$	pK$_a^b$
Arginine	–(CH2)$_3$-NH-C(C = NH)NH$_2$	12.10	12.48
Aspartic acid	–CH$_2$COOH	3.71	3.86
Cysteine	–CH$_2$SH	8.14	8.00
Glutamic acid	–CH$_2$CH$_2$COOH	4.15	4.07
Histidine		6.04[c]	6.10[c]
Lysine	–(CH$_2$)$_4$NH$_2$	10.67	10.53
Tyrosine		10.10	10.07

[a] Lide, D. R., *CRC Handbook of Chemistry and Physics*, Vol. 90, Boca Raton, FL: CRC Press, 2009–2010. With permission.

[b] Parrill, A., *Amino Acid pK$_a$ Values*. With permission. [368]

[c] This value represents dissociation of the protonated imidazolium ring.

6.5 PYRIDINES AND RELATED HETEROCYCLICS

The acidities of substituted pyridines have been studied by a number of groups using several different methods. In 1951 Gero and Markham [371] found a linear relationship between the pK$_a$s of methyl pyridines and the number of methyl substituents. Abboud et al. [372] analyzed the gas-phase and aqueous acidities of a large number of substituted pyridines in terms of polarizability, field, and resonance contributions from the substituents. Chen and MacKerell [353] used AM1 and MP2 calculations to examine substituent effects on pyridine pK$_a$s, concluding that the greatest shortcomings in the calculations stemmed from limitations of the solvent model.

In 1995 Soscún Machado and Hinchliffe [373] used Hartree-Fock 6-31G* and 6-31G** calculations to estimate the pK$_a$s of a set of monocyclic and bicyclic azines. For three classes (see representatives below) they found good correlations (r = 0.91–0.99) between the energies of the highest occupied molecular orbital (HOMO) energies ε_{HOMO} and the pK$_a$s.

Pyridine   Cinnoline   [1,5]Naphthyridine

Tehan et al. [213] applied their semi-empirical AM1 frontier-orbital theory approach to several sets of nitrogen heterocyclic compounds. For a set of 82 pyridines using the nitrogen-atom electrophilic superdelocalizability (SE_1) a single descriptor resulted in $r^2 = 0.57$, whereas a three-term model gave $r^2 = 0.78$. For a set of 13 pyrimidines they found $r^2 = 0.79$ using SE_1 alone, and for 28 quinolines they obtained $r^2 = 0.80$ using the energy of the lowest unoccupied molecular orbital (LUMO) as a single descriptor. For their entire set of 150 heterocyclic compounds they found $r^2 = 0.55$ using SE_1 alone and $r^2 = 0.72$ with three descriptors.

Major, Laxer, and Fischer [374] have used DFT B3LYP/aug-cc-pVTZ(-f) calculations with the PCM solvent model [98] to estimate the pK_as of a collection of 23 heterocyclic compounds, including 7 pyridines. After scaling the atomic radii to achieve good agreement with experimental values they found $r^2 = 0.993$ for the entire heterocyclic set, and the calculated pK_a values for the pyridines are seen to be in good agreement with the reported experimental values.

Habibi-Yangjeh, Pourbasheer, and Danandeh-Jenagharad [375] employed an artificial neural network (ANN) approach, along with principal components analysis (PCA), to analyze the aqueous pK_bs of 91 pyridines. In their approach they used eight principal components as inputs to an 8-12-1 ANN, and stopped the training when the root-mean-square-error (RMSE) started to increase in order to avoid overtraining. They found that the ANN-PCA approach was superior to an alternative multiple linear regression approach, producing the following relationship:

$$pK_b(ANN) = 0.9814 \ pK_b(exp) + 0.1765 \ \text{(pyridines)} \qquad (6.21)$$

$$n = 91, \quad r^2 = 0.984, \quad RMSE = 0.28, \quad F = 5560$$

6.6 PURINES AND PYRIMIDINES

Because of their intimate involvement in the genetic code purines and pyrimidines have received considerable attention from both experimental and theoretical viewpoints. The numbering systems for these compounds are noted below:

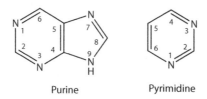

Purine   Pyrimidine

The purines and pyrimidines found in DNA and RNA are shown below:

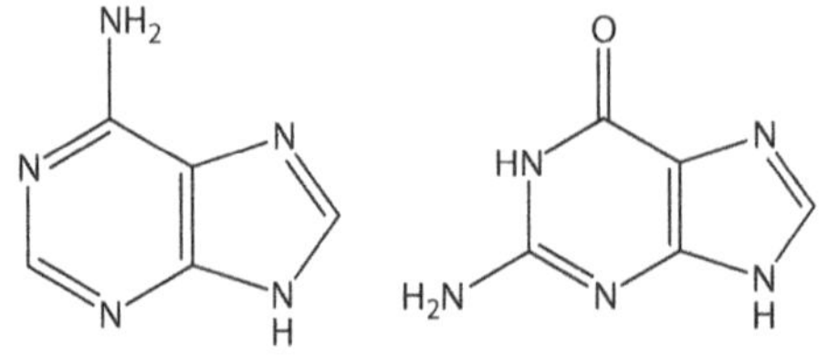

Adenine   Guanine

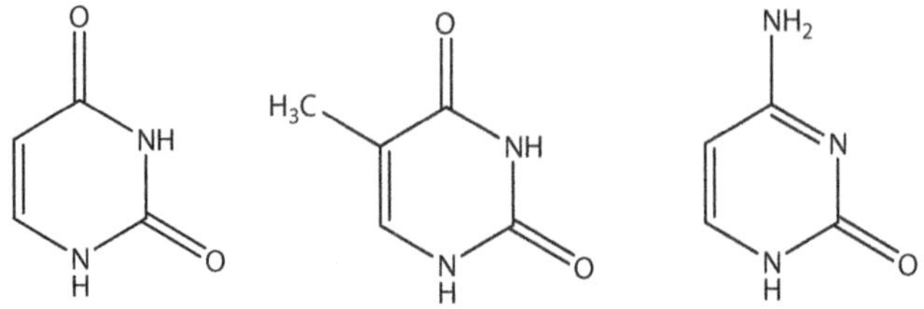

Uracil   Thymine   Cytosine

These compounds can be present in a variety of tautomeric forms both in the gas phase and in solution, and the presence of the proper form is normally crucial for each biological role being played, as for example, in stabilizing the double helix of DNA. Indeed, it was insight into the correct tautomeric forms of the DNA bases that was critical for the discovery of the DNA double-helix structure [376,377]. Moser, Range, and York [301] have provided high-level benchmark calculations of the gas-phase proton affinities and basicities for many of the tautomers of these compounds. A number of additional studies have addressed the tautomeric equilibria of these compounds in both gas and aqueous phases [378–382].

The pK$_a$s of these compounds have been subjects of a large number of experimental and theoretical first-principles investigations. Yang et al. [383] used the B3LYP/aug-cc-pVTZ method with a continuum solvent based on solving the Poisson-Boltzmann equation to estimate the pK$_a$s of 5-substituted uracils. In this way they were able to determine whether dissociation occurred from the N1 or N3 site. Note that 5-methyluracil is thymine and 5-fluorouracil (5-FU) is an important anticancer drug. Whittleton, Hunter, and Wetmore [384] have examined the effects of hydrogen bonding on the acidities of uracil derivatives. Pavel et al. [385] have used Raman spectroscopy and DFT calculations to identify the 5-FU species present at different pHs, and others have examined additional aspects of uracil acidity [386–391].

There have also been a number of theoretical studies of the purines, some of which have already been noted. Major, Laxer, and Fischer [374] have used DFT B3LYP/aug-cc-pVTZ(-f) calculations with the PCM solvent model [98] to predict the pK_as of adenine and 10-substituted adenines. Salter and Chaban [382] have used multi-configurational wave functions to study the gas-phase tautomerism of adenine in its ground and first excited electronic states. Mons et al. [392] have used IR and UV spectroscopy to study the gas-phase tautomers of guanine, and Jang et al. [130] have employed DFT calculations along with a Poisson-Boltzmann continuum solvent model to examine the tautomerism and pK_as of this compound in water.

7 Additional Types of Acids

Several important classes of acids fall outside the classic categories of organic oxyacids and nitrogen acids or have special features that warrant separate discussion. This chapter covers those acids, which include carbon acids, inorganic acids, polyprotic acids, super acids, and excited-state acids.

7.1 CARBON ACIDS

In carbon acids, dissociation takes place from a C-H bond, leading to a negatively charged *carbanion* and a solvated proton or protonated solvent. A number of studies have been devoted to investigations of the carbanions [393,394]. These generally are highly reactive species, although stable carbanions, such as the triphenylmethyl anion, do exist. Carbon acids themselves are normally very weak acids, with pK_a values typically between 20 and 50, but many examples fall well outside this range [395]. Because of measurement difficulties in the high pK_a range, many of the reported experimental pK_a values are highly uncertain, and due to the low aqueous solubilities of many hydrocarbon compounds, pK_as are frequently measured in dimethyl sulfoxide (DMSO) or acetonitrile (AN) solutions.

Conant and Wheland [396] were among the first to investigate the "extremely weak acids" in this class in a systematic way. They chose anhydrous ether as a solvent because of solubility concerns. Of twelve compounds tested, the weakest acid was dimethylphenylmethane, which they estimated to have a pK_a of at minimum 30.5–31. Taking the pK_a of acetophenone as 20, they estimated the pK_a of diphenylmethane at 29.5, with rough estimates for other compounds. This study was followed by McEwen [397], who took methyl alcohol as a standard with $pK_a = 16$. This allowed him to estimate the relative acidities of additional compounds.

The notion of hybridization of the carbon atom has been used to rationalize the pK_a of carbon acids [395,398,399]. The aqueous pK_as of ethane ($pK_a \approx 50$), ethene (ethylene, $pK_a \approx 44$), and ethyne (acetylene, $pK_a \approx 25$)

show a clear trend toward lower pK$_a$ values with lower s-character in the C-H bond [395].

$$CH_4 \quad CH_3\text{-}CH_3 \qquad H_2C=CH_2 \qquad HC\equiv CH$$

$$pK_a \approx 48 \qquad 50 \qquad\qquad 44 \qquad\qquad 25$$

Maksić and Eckert-Maksić [398] found a strong linear relationship between the percent s-character in the C-H bonds and the experimental pK$_a$s of a set of 10 carbon acids:

$$pK_a = 83.12 - 1.33 \text{ (s\%)} \tag{7.1}$$

Given the large uncertainties in the experimental pK$_a$ values for these compounds, some caution must be exercised in interpreting these results, but the overall correspondence is well established.

In 1953 Pearson and Dillon [400] collected a large body of dissociation equilibrium data for what were then called pseudoacids, and discussed the relationship between the dissociation equilibrium constants (K$_a$) and the rate constants (k$_1$) for ionization, noting that the logarithms of the two values were (roughly) linearly related. In 1962 Novikov et al. [401] were able to use spectroscopic methods to estimate the pK$_a$s and temperature dependences of some aliphatic nitro compounds. These were found to be highly acidic: for example, at 20° C the pK$_a$ of CH$_2$(NO$_2$)$_2$ was estimated at 3.60, and that of 1,1-dinitroethane at 5.21. The value for 1,1-dinitropropane was 5.53. In 1973 Breslow and Chu [402] used an electrochemical method to estimate the pK$_a$s of a number of carbon acids. Values for some triphenylmethanes fell in the 30–35 range, while triphenylcyclopropene had a pK$_a$ = 50, and several substituted cyclopropenes had values in the 62–65 range. Later Breslow and Goodin [403] estimated the pK$_a$ of isobutene at 71. Sim, Griller, and Wayner [404] measured the pK$_a$s of toluene and eight substituted toluenes in acetonitrile, finding values ranging from 55.2 for 4-methoxy toluene to 39.8 for 4-C(O)Me-toluene. (The pK$_a$ of toluene itself is 51 in acetonitrile and 42–43 in DMSO.) In 1991 Kresge et al. [405] estimated the pK$_a$s of a number of substituted acetylenes, which fell in the 18–23 range. Streitwieser et al. [406] measured the pK$_a$s of 20 hydrocarbons over the pK$_a$ range of 27–39.

In 1988 Meot-Ner, Liebman, and Kafafi [407] measured the gas-phase acidities of a number of aromatic and heterocyclic carbon acids, and compared these values with those obtained using AM1 [211] calculations. The AM1 calculations were seen to provide reasonable estimates for the deprotonation energies in most cases.

In an early study Jorgensen and Briggs [48] used Monte Carlo simulations and quantum chemical calculations to compute hydration effects on the pK_as of several carbon acids in water. They were able to characterize the hydrogen bonding around the anions in the first solvation shell and arrived at a pK_a value of 52 ± 2 for ethane using this approach. They found $pK_a = 28$ for acetonitrile and $pK_a = 33$ for methylamine.

In 1991 Cioslowski, Mixon, and Fleischmann [408] examined the electronic structures of CHF_3, $CH(CN)_3$, and $CH(NO_2)_3$ at the HF/6-31G* and HF/6-31++G** levels in an effort to understand the high acidities of these species. They were unable to find any simple correlations between the gas-phase acidities and properties of these compounds, such as partial atomic charges and C-H bond lengths. At the same time Murray, Brinck, and Politzer [409] were able to establish correlations between the aqueous pK_as of a set of ten mostly cyclic hydrocarbons and two measures from the molecular electrostatic potential surface, the minimum local ionization energy $I_{S,min}$ and the potential minimum V_{min}.

Alkorta and Elguero [410] used density functional theory (B3LYP/6-311++G**) to examine the gas-phase and aqueous acidities of a set of carbon acids and some inorganic reference acids (HCl, HF, H_2O, H_2). They found an excellent correlation between the experimental and calculated ΔH_{acid} ($= H_R^- - H_{RH}$) values:

$$\Delta H_{acid}(exp) = 21 \ (\pm 7) + 0.95 \ (\pm 0.02) \ \Delta H_{acid}(calc) \qquad (7.2)$$

$$n = 12, \quad r^2 = 0.996$$

The experimental values were taken from March [411], and the energies are in kcal/mol. The aqueous pK_a values of the aliphatic carbon acids were also closely related to both the experimental and calculated ΔH_{acid} values:

$$pK_a = -202 \ (\pm 6) + 0.60 \ (\pm 0.01) \ \Delta H_{acid}(exp) \qquad (7.3a)$$

$$n = 6, \quad r^2 = 0.998$$

$$pK_a = -196 \ (\pm 5) + 0.59 \ (\pm 0.01) \ \Delta H_{acid}(calc) \qquad (7.3b)$$

$$n = 7, \quad r^2 = 0.998$$

These workers also found a close relationship between $\Delta H_{acid}(calc)$ and the percent s-character in the carbon atom hybridization.

In 2000 Topol and co-workers [412] computed the acidities of a number of carbon acids in the gas phase and water using MP2 and density functional theory (DFT) calculations. They examined the use of both a continuum solvent model and explicit waters. For compounds with aqueous pK$_a$s over 40 they found that the addition of explicit waters led to less accurate results.

Charif et al. [413] later carried out a DFT (B3LYP/6-311++G**) study of the gas- and aqueous-phase acidities of a set of 21 carbon acids. They observed a strong correspondence between the experimental aqueous pK$_a$s and the calculated gas-phase Gibbs energy changes of these compounds:

$$pK_a = -135.99 + 0.4453 \ \Delta G°_{gas}(calc) \tag{7.4}$$

$$n = 21, \quad r^2 = 0.969, \quad s = 3.27$$

Note that the close association found between gas-phase and aqueous behavior for these carbon acids is in contrast to the behaviors of alcohols, carboxylic acids, and amines, where solvent effects can sometimes reverse the acidity order. See, for example, the discussion of aliphatic amines in Section 6.1. In this study use of a polarizable continuum solvent model (PCM) actually led to slightly worse results for the calculated aqueous pK$_a$s.

Perez and Perez [414] have discussed the special case of carbon acids where proton dissociation occurs from the central carbon in 1,3-dicarbonyl compounds. In these cases complications arise not only from keto-enol tautomerisms, but also because of geometric flexibility or lack thereof. Notable differences occur between rigid cyclic β-diketones and more flexible acyclic forms, as in the cases of Meldrum's acid (cyclic, pK$_a$ = 7.32 in DMSO)

Meldrum's acid   Dimethyl malonate   Dimedone   2,4-Pentanedione

and dimethyl malonate (acyclic, pK$_a$ = 15.87 in DMSO) and dimedone (cyclic, pK$_a$ = 11.2 in DMSO) and 2,4-pentane dione (acyclic, pK$_a$ = 13.3 in DMSO) [414]. Part of these differences can be explained as stabilization of the enolates in the rigid, cyclic form, while geometrical considerations related to electrostatic and steric interactions appear to account for the remainder.

Richard, Williams, and Gao [415] examined the effect of cyano groups on carbon acidity both experimentally and theoretically. Their experimental pK$_a$ results were CH$_3$CN (28.9), CH$_3$CH$_2$CN (30.9), and NC-(CH$_2$)$_2$-CN (26.6). Theoretically, they employed the semiempirical AM1 method [211] and ab initio methods for the solutes and a Monte Carlo method for the solvent molecules, with special attention to solute-solvent interactions. They noted the progression below for the methane-nitrile series:

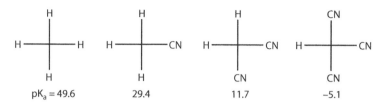

This trend results from the strong electron-withdrawing influence of the cyano ($-C\equiv N$) group. (Note that the early report by Pearson and Dillon [400] gives somewhat different estimates for the pK_a values of these compounds.)

In 2000 Koppel et al. [416] presented an extensive collection of experimental Brønsted acidities of C-H acids in the gas phase and DMSO solution. These authors found that substituent effects in DMSO were attenuated relative to those in the gas phase, but there were strong correlations between the two acidities. Among other results, they estimated the pK_a of toluene as 43.0, with electron-withdrawing substituents on the phenyl ring considerably reducing this value (4-CN-touene, $pK_a = 30.7$; 4-NO$_2$-toluene, $pK_a = 20.4$; 2,3,4,5,6-CN$_5$-touluene, $pK_a = 8.2$; etc.). Direct substitution on the methyl group of toluene is known to dramatically reduce the pK_a ($C_6H_5CH_2CN$, $pK_a = 21.9$; $C_6H_5CH(CN)_2$, $pK_a = 4.2$, in DMSO) [417]. Overall, Koppel et al. provide acidity data for 116 carbon acids.

7.2 INORGANIC ACIDS

The pK_as of some representative inorganic acids are given in Table 7.1. Despite their universal use in industry and commerce, inorganic acids have received far less attention from theoreticians than have the organic acids discussed to this point. This absence may be due to the wide diversity and lack of a common framework for these acids. Whatever the reason, only a very limited number of theoretical attempts have been made to calculate the pK_as of this class.

Most early studies of the strengths of inorganic acids were qualitative, simply comparing acid strengths. In 1938 Kossiakoff and Harker [418] presented one of the earliest attempts at quantitative estimation of the pK_as of inorganic acids, focusing on the prominent group of inorganic oxygen acids. Following several assumptions about features of the ionization process, they obtained an expression for the Gibbs energy of the dissociation process:

$$\Delta G = W_1 - C_1 + RT \ln(n_O/n_H) \tag{7.5}$$

TABLE 7.1

Aqueous pK$_a$s for Some Representative Inorganic Compounds at 25°C[a]

Name	Formula	pK$_a$
Ammonia	NH_3	9.25
Boric acid[b]	H_3BO_3	9.27, >14
Carbonic acid	H_2CO_3	6.35 (3.45[d]), 10.33
Cyanic acid	HCNO	3.46
Hydrazine	N_2H_4	8.1
Hydrocyanic acid	HCN	9.21
Hydrofluoric acid	HF	3.20
Hydrogen peroxide	H_2O_2	11.62
Hydrogen selenide	H_2Se	3.89, 11.0
Hydrogen sulfide	H_2S	7.05, 19
Hydroxylamine	NH_2OH	5.94
Nitrous acid	HNO_2	3.25
Phosphoric acid	H_3PO_4	2.16, 7.21, 12.32
Silicic acid[c]	H_4SiO_4	9.9, 11.8, 12, 12
Sulfuric acid	H_2SO_4	1.99
Sulfurous acid	H_2SO_3	1.85, 7.2
Thiocyanic acid	HSCN	−1.8
Water	H_2O	13.995 (15.74)[e]

[a] Lide, D. R., *CRC Handbook of Chemistry and Physics*, Vol. 90. Boca Raton, FL: CRC Press, 2009–2010.

[b] 20°C.

[c] 30°C.

[d] Adamczyk, K. et al., *Science* 2009, 326, 1690–1694.

[e] Using K$_a$ = [H$^+$][OH$^-$]/[H$_2$O].

In this expression W_1 is an electrostatic energy term for the transfer of a proton from the hydroxyl group to an adjacent solvent molecule, C_1 is a constant term for the solvent, n_O is the number of equivalent non-hydroxyl oxygens, and n_H is the number of transferable hydrogens. For 26 inorganic oxygen acids they found a good correlation with an average deviation of 0.89. In 1948 Ricci [419] critically examined the assumptions employed by Kossiakoff and Harker, simplified the parameters, and presented the following expression for acids of the form H_aMO_b:

$$pK_a = 8.0 - m(9.0) + n(4.0) \tag{7.6}$$

where m is the formal charge of the central atom and n = b − a. This yielded an average deviation of 0.93 for the species studied by Kossiakoff and Harker, and 0.91 for an expanded set of 36 compounds.

In 1983 Bayless [420] modified the above approach to give an empirical equation for the pK_as of nonmetal hydrides of the general form $YH_{(8-G)}$:

$$pK_a = 73.7 - 11.71G - 0.29GH - 378.6(q-1)(1/GP^2) \qquad (7.7)$$

where G is the group in the periodic table of Y, H is the number of hydrogen atoms on Y, P is the period number, and q is the charge on the acid. Taking H_2O as an example, with G = 6, H = 2, P = 2, and q = 0, the value for $pK_a(H_2O)$ is 15.74, in exact agreement with the experimental value of 15.74. This equation gives an excellent account, with an average deviation of just 0.1 pK_a units, for a well-characterized set of five compounds (H_2O, H_2S, H_2Se, H_2Te, and HF), and a good account of the experimental pK_as for 12 additional, less well-characterized nonmetal halides.

The hydrohalic acids (HF, HCl, HBr, and HI) form an important and informative subclass of inorganic acids. The acidities of these acids decrease in the order HI > HBr > HCl >> HF. The pK_a of HF is firmly established, but there is disagreement regarding the specific pK_as for the other three acids (see Table 7.2). This disagreement arises because of the difficulty of measuring pK_as at the extreme ends of the range, in this case at negative values (and for the carbon acids considered earlier, at high pK_a values). Because of such experimental limitations the pK_a values for this region are typically estimated using a variety of assumptions, some of which may be less firmly based than others. Hence a range of values has been reported for the higher halides in this series. (Because of the solvent-related shifts in pK_a, measurements in DMSO are more tractable, and estimates yield $pK_a = 15 \pm 2$ for HF [421], 1.8 for HCl [422], and 0.9 for HBr [422].)

In addition, it is of interest to consider why HF ($pK_a \approx 3.20$) is a relatively weak acid in this series, whereas the others are strong acids, all with negative pK_as. Pauling [423] addressed this point and explained the separation between HF and the other haloacids in terms of the electronegativity differences between the halogens and hydrogen. He pointed out that a similar variation is found in the series H_2O ($pK_a = 14$), H_2S (7.05), H_2Se (3.89), and H_2Te (2.6). At about the same time as Pauling's study McCoubrey [424] presented a similar analysis. Later Myers [425] examined the hydrogen halide series using different assumptions and arrived at different values for the pK_as of the higher halogen acids, drawing some criticism from Pauling [426,427]. In 2001 Schmid and Miah [428] presented a different analysis using simple solvation models.

Frigden [429] has addressed the underlying causes of the differing hydrohalide acidities in terms of the physical properties of the compounds. As shown in Table 7.2, several properties such as the trend to lower bond dissociation energies (BDEs) and smaller differences in X-H electronegativities ($\Delta\eta$) as one proceeds down a group in the periodic table are consistent

TABLE 7.2
Some Properties of the Hydrohalic Acids

Acid	R(H-X) pm	BDE[a] kJ/mol	Dipole (D)	$\Delta\eta$[b]	pK$_a$[c]	pK$_a$[d]	pK$_a$[e]	pK$_a$[f]	pK$_a$[g]	pK$_a$[h]	pK$_a$[i]
HF	91.7	570	1.8262	1.9	3.20	3.1	3.15	3.18	3.2	2.4	3
HCl	127.4	431	1.1086	0.9	-3	-3.9	-1.8	-7	-6.3	-7	-7
HBr	141.4	366	0.8272	0.7	-6	-5.8	-4.7	-9	-8.7	-9	-9
HI	160.9	298	0.448	0.4	-7	-10.4	-5.2	-11	-9.3	-9.5	-10

[a] Homolytic bond dissociation energy. From Lide, D. R., *CRC Handbook of Chemistry and Physics*, Vol. 90. Boca Raton, FL: CRC Press, 2009–2010.

[b] Electronegativity difference of atoms. From Fridgen, T. D., *Journal of Chemical Education* 2008, 85 (9), 1220–1221.

[c] Fridgen, T. D., *Journal of Chemical Education* 2008, 85 (9), 1220–1221.

[d] Schmid, R., and Miah, A. M., *Journal of Chemical Education* 2001, 78 (1), 116–117.

[e] Myers, R. T., *Journal of Chemical Education* 1976, 53 (1), 17–19.

[f] Albert, A., and Serjeant, E. P., *Ionization Constants of Acids and Bases*. London: Methuen, 1962.

[g] Pauling, L., *Journal of Chemical Education* 1956, 33, 16–17.

[h] McCoubrey, J. C., *Transactions of the Faraday Society* 1955, 51, 743–747.

[i] Bell, R. P., *The Proton in Chemistry*, 2nd ed., Ithaca, NY: Cornell University Press, 1973.

with the order of the aqueous pK_as for these compounds. This same trend is evident in the series H_2O, H_2S, H_2Se, and H_2Te. Unfortunately, as Frigden also notes, just the opposite trend is found if one proceeds across a period of the periodic table, e.g., in the series $H-CH_3$, $H-NH_2$, $H-OH$, and $H-F$.

Several experimental studies have examined sulfamic acid (H_3NSO_3), which exists predominantly in its zwitterion form in its crystals and in its neutral form in aqueous solution, and is described as the "simplest amino sulfonic acid" [430]:

In 1951 Taylor, Desch, and Catotti [431] estimated the aqueous pK_a of sulfamic acid as 1.0 using conductance measurements. Subsequently King and King found a pK_a value of 0.986 [430] using an EMF technique. These latter workers were also able to estimate the thermodynamic functions $\Delta G°$, $\Delta H°$, $\Delta S°$, and ΔC_p from the temperature dependence of the dissociation constant. More recently, Gelb and Alper [279] have investigated the acidity of sulfamic acid and some haloacetic acids using potentiometric as well as conductometric methods. For sulfamic acid at 20°C they found $pK_a \approx 1.02$, in agreement with earlier studies, but they found a divergence in the results from the two experimental methods for the haloacetic acids, which they attributed to the presence of an ion-associated (H^+A^-) species contributing to the electrolytic conductance.

In some cases a small number of inorganic acid pK_as have been estimated along with organic acid pK_as, most often in a collection of first-principles calculations [79,193,410,432,433]. In other cases specific inorganic compounds, e.g., nitrous acid (HNO_2), individual compounds are focused upon [77]. For HNO_2 da Silva and coworkers [77] measured the aqueous pK_a at 3.16 and determined the enthalpic ($\Delta H° = 6.7$ kJ/mol) and entropic ($\Delta S° = -38.4$ J/mol-K) contributions to the Gibbs energy of the dissociation from the temperature dependence. These workers used a variety of computational approaches to determine the gas-phase acidities (experimental value = 1396.2 kJ/mol) and aqueous pK_as for the *trans* and *cis* forms of the acid. Two of the DFT methods were able to estimate the aqueous pK_a to within ±0.2 pK_a units.

7.3   POLYPROTIC ACIDS

Compounds with more than one dissociating group can pose some complications, especially when they contain equivalent groups or dissociable groups with close-lying pK_a values [93,173,434]. Bouchoux [356,435] has reviewed the gas-phase basicities of polyfunctional molecules in some detail.

In the absence of equivalent groups we have already seen in Chapter 6 that for amino acids at normal (~7) pH both the carboxylic and amino groups are ionized, forming a zwitterion. Alternatively, in a compound there might be several possible sites of ionization of the same functional group. For example, in ascorbic acid (vitamin C, pK_a = 4.1), shown below, there are four different –OH groups, and it is of interest to determine which of these groups will be the first to ionize. Because aliphatic alcohols have pK_as near 16, we can safely suppose that neither hydroxyl 1 or 2 will easily ionize. A simple calculation of the energies of the anions formed by dissociations at hydroxyls 3 and 4 shows that the 3-position anion is the more stable, with lower energy, and is the site responsible for the first ionization [21].

A further example is the well-known chemiluminescent compound *luminol* [436]: Luminol reacts with the iron in hemoglobin and is used in forensics to detect traces of blood; its pK_as are reported to be 6.74 and 15.1 [436]. The first ionization takes place at the lower of the two ring N-H groups.

Figure 7.1 shows the pK_as of some dicarboxylic acids along with the pK_as of some related monocarboxylic acids. As noted earlier, complications can arise when a compound contains groups that have dissociations with close pK_a values [93,173,434]. In such a case one must consider the equilibria shown in Figure 7.2 for dissociation of an original acid RH_aH_b. [173]. In this scheme K_A, K_B, K_C, and K_D are so-called microscopic equilibrium constants and K_T is the equilibrium constant for tautomerization between RH_{a-} and RH_{b-}. The experimentally measured (macroscopic) dissociation constants K_1 and K_2 pertain to the following equilibria:

$$K_1: \quad RH_aH_b \rightleftharpoons RH_b^- \text{ or } RH_a^- + H^+$$

$$K_2: \quad RH_a^- \text{ or } RH_b^- \leftrightharpoons R^{2-} + H^+$$

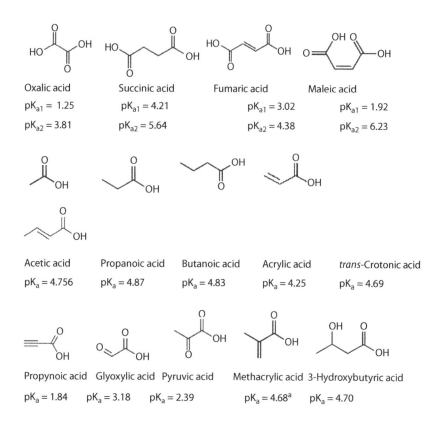

Oxalic acid
$pK_{a1} = 1.25$
$pK_{a2} = 3.81$

Succinic acid
$pK_{a1} = 4.21$
$pK_{a2} = 5.64$

Fumaric acid
$pK_{a1} = 3.02$
$pK_{a2} = 4.38$

Maleic acid
$pK_{a1} = 1.92$
$pK_{a2} = 6.23$

Acetic acid
$pK_a = 4.756$

Propanoic acid
$pK_a = 4.87$

Butanoic acid
$pK_a = 4.83$

Acrylic acid
$pK_a = 4.25$

trans-Crotonic acid
$pK_a = 4.69$

Propynoic acid
$pK_a = 1.84$

Glyoxylic acid
$pK_a = 3.18$

Pyruvic acid
$pK_a = 2.39$

Methacrylic acid
$pK_a = 4.68^a$

3-Hydroxybutyric acid
$pK_a = 4.70$

FIGURE 7.1 Some diprotic carboxylic acids and related monocarboxylic acids. (From Volgger, D. et al., *Journal of Chromatography A* 1997, 758, 263–276. With permission.)

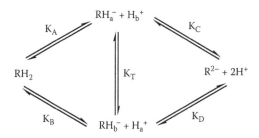

FIGURE 7.2 Equilibria for close-lying values of pK_a. (After Perrin, D. D., et al., *pK_a Prediction for Organic Acids and Bases*, New York: Chapman & Hall, 1981. With permission.)

There are several relationships between the constants, the most important being:

$$K_1 = K_A + K_B$$
$$1/K_2 = 1/K_C + 1/K_D$$

Examples of applications of this analysis include the dissociations of aminobenzoic acids [437], arginine [438], the anti-inflammatory drug niflumic acid [439], and the antihistamine cetirizine [440].

In some cases, as for the amino acids where K_T strongly favors the zwitterion form, the path $^+H_3N\text{-}R\text{-}COOH \rightarrow \ ^+H_3N\text{-}R\text{-}COO^- + H^+ \rightarrow H_2N\text{-}R\text{-}COO^- + 2H^+$ dominates and the alternative path can be neglected. In this case the microscopic constants K_A and K_C can be used to approximate pK$_1$ and pK$_2$ with little loss of accuracy.

A further caution must be exercised when a compound possesses two or more *equivalent* dissociating groups. For statistical reasons, in the case of N equivalent sites the measured pK$_a$ will be less by logN than the pK$_a$ of an analogous monofuntional acid [173]. This statistical correction arises because there are N ways for such an acid to dissociate to form its mono-anion, but only one way for the monoanion to undergo the reverse process. For example, succinic acid (Figure 7.1) has two equivalent –COOH groups [173] and a pK$_{a1}$ $\approx$ 4.21, whereas butanoic acid has pK$_a$ $\approx$ 4.83. For pK$_{a1}$ the statistical correction of log2 = 0.30 brings the estimated pK$_{a1}$ for succinic acid to 4.83 – 0.30 = 4.53. Conversely, the statistical correction for the second dissociation should lead to a value 0.3 *greater* than the mono-acid pK$_a$ value, placing the estimated pK$_{a2}$ at $\approx$ 5.13. The observed pK$_{a2}$ of succinic acid is 5.64. Clearly one must always be aware of additional influences affecting the individual acids, such as inductive, resonance, and steric effects, and hydrogen bonding.

7.4   SUPERACIDS

Superacids, which can be 10^7 to 10^{10} times stronger than 100% sulfuric acid, have attracted considerable interest in the past several decades due to their increasing use as catalysts for synthesis and their fundamental importance in organic research [441]. The term *superacid* was first employed by Hall and Conant in 1927, referring to highly acidic solutions in nonaqueous media [442]. Later the term came to be applied to solutions more acidic than 100% sulfuric acid (Hammett acidity function H_0 = –12) [443,444]. Examples of Brønsted acids more acidic than sulfuric acid include fluorosulfuric acid (HSO$_3$F, H_0 $\approx$ –15.1) and trifluoromethanesulfonic acid (CF$_3$SO$_3$H, H_0 $\approx$ –14.1) [441,445]. With Lewis acids, Olah, Prakash, and

TABLE 7.3

Properties of Some Brønsted and Lewis Superacids [441]

Acid	Melting Point (°C)	Boiling Point (°C)	Density (g/cm³)	Dielectric Constant	Hammett Acidity (H_0)
Brønsted Acids					
$HClO_4$	−112	110 (explodes)	1.767	—	~13.0
HSO_3Cl	−81	151 (decomposes)	1.753	60 ± 10	13.8
HSO_3F	−89	162.7	1.726	120	15.1
CF_3SO_3H	−34	162	1.698	—	14.1
HF (neat)	−83	20	0.698	84	15.1
Lewis Acids					
SbF_5	7	142.7	3.145[a]	—	—
AsF_5	−79.8	−52.8	2.33[b]	—	—
TaF_5	97	229	3.9[a]	—	—
NbF_5	72–73	236	2.7[a]	—	—

[a] At 15°C.
[b] At the BP.

Sommer suggested that anhydrous aluminum chloride be taken as a reference, with more acidic examples being referred to as superacids [446]. The properties of some representative superacids are summarized in Table 7.3.

Fluorosulfuric acid   Trifluoromethanesulfonic acid

Even stronger media can be produced by adding a strong Brønsted or Lewis acid (or both) to an already strong acid. In the late 1960s Olah and Schlosberg produced "magic acid" [447], a combination of fluorosulfuric acid and antimony pentafluoride (SbF_5) that was capable of protonating hydrocarbons.

Recently Kütt and co-workers have provided an extensive compilation of the equilibrium acidities of molecular superacids in 1,2-dichloroethane (DCE) [448]. DCE has weak basic properties and high polarity and is a favorable solvent for measuring the acidities of especially strong acids. The authors list the pK_a values for 62 superacids in DCE, including all of the common strong mineral acids, on a relative scale spanning 15 orders of magnitude, and running from picric acid ($pK_a = 0$ to establish the scale) to 1,1,2,3,3-pentacyanopropene. Although the correlation between the DCE scale values and gas-phase acidities was poor, the authors found a good correlation between their DCE values and pK_as in acetonitrile, allowing them to estimate pK$_a$ values in the latter solvent. Trummal et al. [449] have performed SMD/M05-2X/6-31G* calculations on some CH and NH superacids in DCE and report a mean unsigned error of 0.5 pK$_a$ units, with $r^2 = 0.990$.

7.5   EXCITED-STATE ACIDS

The science of photochemistry is based on the fact that elevation of a molecule to an excited state—normally carried out through absorption of a photon—creates a new chemical species with a changed electronic configuration and often a new geometry. The excited species created in this way will necessarily undergo reactions that differ to some extent from those that dominate in the ground state. Among the affected reactions it is often found that the tendency of the compound to undergo acid dissociation is altered in the excited state, and may be either increased or diminished.

Among the earliest observations of this phenomenon was the demonstration by Förster [450]—following up an earlier observation by Weber—that acidic aqueous solutions of the dye fluorescein, which exhibit just a single species (the cation) in absorption, showed fluorescence from both the cation and an additional, dissociated species. It was concluded that the excited fluorescein cation was more acidic than its ground-state counterpart. Later work supported this conclusion and supplied additional examples of altered acidities and basicities in excited states [451–453].

Early reviews on the acid-base properties of excited states were presented by Weller [451], Vander Donckt [454], Schulman and Winefordner [455], Schulman [456], and Ireland and Wyatt [457]. Discussions of this subject also appeared in the influential books by Förster [458] and Parker [459]. More recently, Agmon has reviewed the elementary steps in excited-state proton transfer [460].

In 1965 Jaffé and Jones [461] applied the Hammett equation to values of excited-state pK$_a$s and found a rough correlation with ground-state σ values. That same year Wehry and Rogers [462] showed that ground-state Hammett and Taft constants could be applied to the excited singlet and triplet states of phenol and 16 monosubstituted phenols to produce a satisfactory correlation.

In 1966 Stryer [463] and Wehry and Rogers [464] reported deuterium effects on acid dissociation in excited states. Stryer studied a collection of naphthalene derivatives, phenols, and indoles, and concluded that both the forward and reverse reaction rates were slower in D$_2$O than in H$_2$O. Wehry and Rogers examined a series of phenols and aromatic carboxylic acids and found that for the phenols the pK$_a$s in the excited singlet states were about 6–7 pK$_a$ units lower than in the corresponding ground states. In contrast, the excited-state pK$_a$s of benzoic and naphthoic acids were found to be about 5–6 pK$_a$ units *higher* than those in their ground states. Shortly after this Vander Donckt and Porter [465] issued a corrective to the carboxylic acid values, and added results for 1-, 2-, and 9-anthroic acids, indicating smaller pK$_a$ increases (ca. 2–4 pK$_a$ units) in going from the S$_0$ to the S$_1$ states. A further caution is that results for excited-state pK$_a$s obtained from different methods can vary considerably [466–468].

1-Naphthol is more acidic by about 2 pK$_a$ units in its excited singlet state than is 2-napthol. Tolbert and Haubrich [469,470] have studied the excited-state pK$_a$s of cyanonaphthols, utilizing the electron-withdrawing power of the –CN group to increase the acidity of the compounds. They showed that the resulting enhancement of acidity in the excited states allowed excited-state proton transfer to be observed in both alcohol solutions and DMSO. This observation demonstrated that a Grotthuss mechanism [471–474] (H$^+$ transfer via surrounding waters) was not necessary for the *excited-state proton transfer* (ESPT).

Pines and Huppert have utilized the enhanced excited-state acidity of 8-hydroxy-1,3,6-pyrenesufonate (HPTS) to design a laser-induced "pH jump method" [475]. HPTS has a pK$_a$ of 7.8 ± 0.1 in its ground state and a pK$_a$ of 0.4 ± 0.1 in its first excited singlet state. In this method a laser pulse excites HPTS to its excited state, which quickly dissociates and causes a jump in the acidity of the solution. The consequences of this spike in acidity can be followed as the system then relaxes to its normal state. For example, this method has allowed Adamczyk et al. [476] to establish that carbonic acid (H$_2$CO$_3$) can exist as a stable entity in solution and to estimate its pK$_a$ at 3.45 ± 0.15, in contrast to earlier estimates of 6.35. Tolbert and Solntsev [468] have used general principles of substituent effects on excited states to design "super" *photoacids* that allow revealing studies of a host of ultrafast reactions. As these authors note, the role of tunneling in the excited-state dissociations remains to be clarified.

8 Acidities in Nonaqueous Solvents

Although the great majority of pK_a measurements have been carried out in aqueous solutions, a significant number of measurements have been made in other solvents, and findings in these solvents can be highly informative regarding the inherent nature of the dissociation process as well as of practical importance because of the common use of nonaqueous solvents in organic chemistry and industry. How the solvent environment modifies the intrinsic gas-phase acid dissociation is a feature of fundamental importance, and can lead to a broader understanding of the ionization process.

When an acid dissociates, as expressed by the relation $HA \rightleftharpoons A^- + H^+$, a major concern is the extent to which the solvent stabilizes the species involved, and in particular the ionic species A^- and H^+, since these species are normally more strongly affected. This stabilization will depend on both the nature of the solute and the characteristics of the solvent, such as its dielectric constant and whether it is protic or nonprotic. Solvents that stabilize ions poorly will drive the equilibrium to the left, decreasing the acidity and increasing the pK_a. Because water is an exceptionally good solvent for ionic species, the pK_as observed in almost all other solvents are higher than those in water. A list of several properties of water and some common nonaqueous solvents is given in Table 8.1.

8.1 DEUTERIUM OXIDE

As a solvent deuterium oxide ("heavy water," D_2O), although not specifically a nonaqueous solvent, provides some interesting contrasts to the closely related H_2O. Table 8.2 compares several properties of the two solvents.

Significant amounts of deuterium oxide for research first became available in the early 1930s. Apparently the first study to compare acidities in H_2O and D_2O was by Lewis and Schutz [480], who determined the pK_as of acetic acid and chloroacetic acid in H_2O and D_2O. For acetic acid they found $pK_a(CH_3COOH) = 4.74$ and $pK_a(CH_3COOD) = 5.35$, and for chloroacetic acid they found $pK_a(CH_2ClCOOH) = 2.70$ and $pK_a(CH_2ClCOOD) = 3.20$, values close to the presently accepted values. They discussed the differences in terms of the difference in zero-point energies, which should make the energy needed to remove a deuteron greater than that for a proton, and

TABLE 8.1
Properties of Water and Some Nonaqueous Solvents

Solvent	Formula	Abbreviation	Melting Point, °C	Boiling Point, °C	Dielectric Constant ε (20°C)	Viscosity (mPa-s)	Protic
Water	H$_2$O	—	0.0	100.0	80.1	0.890	Yes
Dimethyl sulfoxide	(CH$_3$)$_2$S = O	DMSO	17.89	189	47.24	1.987	No
Acetonitrile	CH$_3$CN	AcCN	−43.82	81.65	36.64	0.369	No
Methanol	CH$_3$OH	MeOH	−97.53	64.6	33.0	0.544	Yes
Ethanol	CH$_3$CH$_2$OH	EtOH	−114.14	78.29	25.3	1.074	Yes
1,2-Dichloroethane	C$_2$H$_4$Cl$_2$	DCE	−35.7	83.5	10.42	0.779	No
Tetrahydrofuran	C$_4$H$_8$O	THF	−108.44	65	7.52	0.456	No

TABLE 8.2

Comparison of Some Properties of H_2O and D_2O

Property	H_2O	D_2O
Freezing point (°C)	0.0	3.82
Boiling point (°C)	100.0	101.4
Density at STP (g/cc)	0.9982	1.1056
Maximum density temperature (°C)	4.0	11.6
Viscosity at 20°C (mPa-s)	1.0016	1.2467
Heat of vaporization (cal/mol)	10,515	10,864
pH at 25°C	6.9996	7.43

in terms of separation of the ions from a weak acid in a solvent. Additional experiments in the 1930s confirmed the general conclusion that weak acids are somewhat more acidic in H_2O than in D_2O [481,482].

In 1959 Ballinger and Long [483] made a careful study of the dissociation constant of trifluoroethanol in H_2O and D_2O, and found $K_a(H_2O) = 4.3 \times 10^{-13}$ and $K_a(D_2O) = 0.95 \times 10^{-13}$, so that the respective pK_as were 12.4 and 13.0. This turns out to be a rather typical result, at least for alcohols, where the acidity constant K_a in H_2O is ca. 3–5 times greater than that in D_2O, and $pK_a(H_2O)$ is roughly 0.6 pK_a units lower than $pK_a(D_2O)$. Bunton and Shiner [484] concluded that the D_2O effect was due to zero-point energy changes mainly associated with hydrogen bonds. In 1963 Bell and Kuhn [485] determined the pK_as of a number of alcohols, phenols, and carboxylic acids in H_2O and D_2O and obtained some evidence that $\Delta pK_a = pK_a(D_2O) - pK_a(H_2O)$ increased with $pK_a(H_2O)$ for alcohols and phenols, but no such evidence for carboxylic acids.

In 1966 Wehry and Rogers [464] examined the pK_as of 2-naphthol and several substituted phenols and found $\Delta pK_a = pK_a(D_2O) - pK_a(H_2O)$ ranged from 0.48 to 0.70. The same general trend held true for pK_as in the lowest excited singlet states of the phenols and for the pK_as in a small set of benzoic acids and naphthoic acids. Stryer [463] found that excited-state proton transfer rates in a set of aromatics were greater in H_2O than in D_2O. As a result, the fluorescence quantum yields were higher in D_2O than in H_2O in most cases. Jencks and Salvesen [486] found similar effects in a small group of thiol acids, with $\Delta pK_a = 2.0–2.5$.

Tremaine and co-workers [487,488] have studied the variations of the ionization constants of 2-naphthol, boric acid, phosphoric acid, and acetic acid in H_2O and D_2O at high temperatures. The pK_as of 2-naphthol, phosphoric acid, and boric acid passed through minima and then increased as the temperature increased from 25°C to 300°C. The ΔpK_a values varied somewhat with temperature. For acetic acid the pK$_a$ increased with temperature and ΔpK$_a$ varied from 0.425 to 0.443.

8.2 DIMETHYL SULFOXIDE

The most frequently used nonaqueous solvent for pK$_a$ measurements has been dimethyl sulfoxide (DMSO, Me$_2$S=O), in large part due to the extensive work of Bordwell and his coworkers [421,489–491]. DMSO is an exceptional solvent: a strong hydrogen bond acceptor, polar, aprotic, and possessing a relatively high dielectric constant. Taft and Bordwell have presented a detailed analysis of the factors, both intrinsic (structure based) and extrinsic (solvent induced), influencing the dissociation of an acid in DMSO, as revealed by differences between the gas-phase and DMSO acidities of 76 diverse acids [417]. Influential intrinsic features include a positive charge on the acid (NH$_4^+$, Ph-NH$_3^+$, where Ph = phenyl), the electronegativity of the central atom (as shown by the acidity order CH$_4$ < NH$_3$ < OH$_2$ < HF), charge delocalization in the anion A$^-$, anion stabilization due to phenyl substitution for a hydrogen [492], ring closure giving rise to enforced planarity (increasing acidity), and steric inhibition of planarity (decreasing acidity), whereas extrinsic effects include solvation of charge-localized anions, decrease of lone pair repulsions by solvation, and attenuation of dipolar substituent effects by the solvent.

Clark and co-workers [493] have used DFT and MP2 computations along with natural bond orbital (NBO) analysis and features of the molecular electrostatic potential (MEP) surface (EPS) to explain why DMSO is such a good solvent. The MEP study revealed a number of strongly positive and negative regions that encourage intermolecular interactions, while the NBO analysis showed that the supposed S=O double bond in (CH$_3$)$_2$SO is actually a coordinate covalent S$^+$→O$^-$ single bond.

In 1975 McCallum and Pethybridge [422] examined the behavior of strong acids, such as methane sulfonic, sulfuric, and nitric acids, in DMSO and concluded that these acids can be considered to be completely dissociated in dilute solution. Farrell, Terrier, and Schaal [267] later compared the pK$_a$s of nitroaromatics in both water and DMSO to examine the effects of solvation on these compounds. For the most part the pK$_a$s of the nitro-phenols, -anilines, -toluenes, and -diphenylamines were slightly higher in DMSO than in water, but in some cases this order was reversed. These authors concluded that these compounds undergo only very limited *specific* solvation in both solvents, and the charges in their anions are largely delocalized.

Koppel and co-workers [416] measured the gas-phase and DMSO acidities of a large number of neutral carbon acids, and found that in general the acidities of these compounds were attenuated by transfer to the DMSO solution. They found rather strong gas phase–DMSO correlations within different classes of compounds, as represented by equations of the form

$$\Delta G_{acid}(\text{gas phase}) = a + b \, pK_a(\text{DMSO}) \tag{8.1}$$

Correlation coefficients for the different classes studied ranged from r = 0.888 to r = 0.999.

Pliego and Riveros [494] obtained the Gibbs energies of solvation for a large number of mineral and organic acids and ions in both aqueous solution and DMSO based on published pK_a values. They concluded that the ions are much better solvated in water than in DMSO. The Gibbs energy values obtained should provide useful benchmarks for the development of continuum solvation models. (But note that these authors used the value $\Delta G°_{solv}(H^+) = -264.0$ kcal/mol for aqueous solution, rather than the more recent value -265.9 kcal/mol recommended by Kelly, Cramer, and Truhlar [55].)

Fu et al. [79] have used MP2 and DFT calculations to estimate the gas-phase acidities and DMSO pK_as of a diverse set of 105 organic acids. The gas-phase acidities were estimated with a precision of 2.2–2.3 kcal/mol and the DMSO pK_as to within 1.7–1.8 pK_a units. These workers later developed a protocol that estimated the pK_as of more than 250 structurally unrelated compounds in DMSO to within ± 1.4 pK_a units [495]. They were able to employ this protocol to estimate redox potentials and bond dissociation enthalpies for the compounds. This team also examined the pK_a values of C-H bonds in aromatic heterocyclic compounds in DMSO, reaching a precision of ± 1.1 pK_a units [496].

In their analysis of organic acids without traditional carboxylic acid groups Perez and Perez [414] have discussed the pK_as of Meldrum's acid and its analogs in DMSO. This example illustrates the acidity-enhancing role of the rigid structure found in Meldrum's acid ($pK_a = 7.32$), as contrasted to the acidity of its acyclic, flexible analog dimethyl malonate ($pK_a = 15.87$).

Meldrum's acid Dimethyl malonate

Gao [497] has presented an insightful analysis comparing the acidities of water and methanol in the gas phase, aqueous solution, and DMSO. In the intrinsic gas-phase reaction methanol ($\Delta G^* = 1569.4$ kJ/mol, where * indicates a standard state of 1 mol/L) is 36.0 kJ/mol more acidic than water ($\Delta G^* = 1605.4$ kJ/mol). This can be understood in terms of greater stabilization found in the CH_3O^- anion compared to the OH^- anion through greater delocalization of the negative charge. In aqueous solution, however, methanol (pK$_a$ = 15.5) is only very slightly more acidic than water (pK$_a$ = 15.7), in this case because of the much stronger solvation of the OH^- ion than of the more charge-delocalized CH_3O^- anion. In DMSO—where pK$_a$(MeOH) = 29.0 and pK$_a$(H_2O) = 31.4—both compounds are far less acidic than in aqueous solution. In fact, as Gao notes, methanol is less acidic in DMSO than in water by a factor of roughly 10^{14}. This dramatic difference results largely from the much weaker solvation of the dissociation product anions by the aprotic DMSO solvent. Gao also notes that calculations using explicit waters can account for a good portion of the ion-water interaction energies in aqueous solution.

More recently Zhu, Wang, and Liang [498] have used DFT and PCM cluster-continuum computations to estimate oxidation potentials, bond dissociation energies, and pK$_a$s of 118 hydroquinones and catechols in DMSO. The pK$_a$s and other properties were also found to correlate well with the Hammett substituent constants for the compounds.

8.3   ACETONITRILE

Acetonitrile, CH_3CN, is an aprotic, weakly basic solvent with a moderate dielectric constant (ε = 36.64) that allows separation of ion pairs to free ions. Moreover, it is transparent in the ultraviolet region, allowing the use of spectroscopic pK$_a$ measurements.

As early as 1933 Kilpatrick and Kilpatrick [499] were able to compare the relative acid strengths of a number of compounds in acetonitrile (AN) and to compare that order with the order of the pK$_a$s in water. One of the first quantitative determinations was a study by H.K. Hall Jr. in 1957 [190] that examined steric effects on the basicities of cyclic amines in water and AN. This study yielded the pK$_a$s of 15 methyl piperidines and some related amines at 30 and 59.6°C, plus information on the thermodynamic parameters. Additional early acidity studies in AN were carried out by Kolthoff and his co-workers [500–502] and by Coetzee and Padmanabhan [503].

In 2006 Kütt and co-workers [504] reported the first comprehensive spectrophotometric acidity scale for AN, including 93 acids spanning 24 orders of magnitude in acidity. The pK$_a$ scale ranged from 3.7 (for 4-Cl-C_6H_4-SO(NTf)NHSO$_2$C$_6$H$_4$-4-NO$_2$, where Tf = CF$_3$SO$_2$–) to 28.1 (for 9-C_6F_5-fluorene). The scale was cross-validated by relative acidity measurements, at least two for every data point. These workers also reported corresponding pK$_a$ values in other solvents such as DMSO, H_2O, heptane, and 1,2-dimethoxy-ethane (DME).

In 2007 Rõõm et al. [505] reported measurements of acidities of diamines and related compounds in the gas phase, AN, and tetrahydrofuran (THF), along with DFT B3LYP/6-311+G** calculations for the gas-phase acidities. They found that the basicity trends in the solvents were different from those in the gas phase, and considerably reduced in scope. They also found that solvent effects in AN were generally intermediate between those in the gas phase and those in THF. They compared their results with literature values for aqueous solutions, and found that among the solvents the pK_a values typically fell in the order H_2O < THF < AN. For example, the pK_as for aniline in water, THF, and AN were 4.6, 5.2, and 10.62, respectively, and those for 1,3-ethane-dimethylaminodiamine were 9.15, 12.8, and 18.68, respectively.

A number of studies—both experimental and theoretical—have examined particular classes of compounds in AN. Barbosa, Beltrán, and Sanz-Nebot studied the pK_as of pH reference materials [506] and quinolone antibacterials [507] in AN-water mixtures. Berdys et al. [508] studied substituted 4-nitropyridine N-oxides in AN. Li et al. [509] examined the basicities of amines and phosphines in AN, and Şanli et al. [510] studied substituted sulfonamides in AN-water mixtures, which they correlated with Kamlet's solvatochromic parameters [511]. The authors note the presence of preferential solvation within the solvent mixtures.

Eckert-Maksić et al. [512] studied the acidities of some guanidines in AN and performed DFT calculations to estimate the gas-phase acidities. The measured $pK_a(AN)$ values ranged from 24.7 to 27.2. Eckert and co-workers [513] applied the COSMO-RS solvent model to a diverse collection of 93 organic acids in AN solution, with good results ($r^2 = 0.97$ and s = 1.38 pK_a units). The results show that AN has only a poor ability to solvate anions. Glasovac, Eckert-Maksić, and Maksić [514] examined the basicities of a large number of bases and superbases in AN using DFT and a polarized continuum solvent model.

8.4   TETRAHYDROFURAN

Tetrahydrofuran is a widely used industrial solvent. It is aprotic, with a low dielectric constant (7.52), low viscosity, and a fairly wide liquid range. Many of the early measurements of acidities in THF were made by the research group of Robert Fraser at the University of Ottawa in Canada. These workers measured the pK_as of weak carbon acids [515], monosubstituted benzenes [516], secondary amines [517], pyridines [518], and heteroaromatic carbon acids [519] using ^{13}C nuclear magnetic resonance (NMR) measurements in the period 1983–1985. Additional pK_a values in THF were determined by Streitwieser et al. [520].

Tetrahydrofuran

Rõõm et al. measured the basicities (in terms of the pK$_a$ values of the conjugate acids HB$^+$ of the bases B) of alkanediamines in gas phase, acetonitrile (AN), and THF, and computed gas-phase basicities for a number of monoamines and diamines using B3LYP/6-311+G** calculations [505]. They examined the intrinsic acidities of the compounds with attention to internal hydrogen bonding and steric strains. In THF, because of its low dielectric constant, they regarded the ions HB$^+$ and A$^-$ to be fully ion-paired. For the most part the pK$_a$s in THF were intermediate to those in AN and H$_2$O: for triethylamine, for example, the pK$_a$s were 18.82 (AN), 12.5 (THF), and 10.7 (H$_2$O).

8.5   1,2-DICHLOROETHANE

1,2-Dichloroethane (DCE) is a reasonable choice for studies of strongly acidic species because it is inert, only very weakly basic, and its permittivity (10.42) is sufficiently low that ion pairs do not dissociate to free ions (see Section 8.6). Accordingly, in 2011 Kütt and coworkers [448] presented a comprehensive collection of acidities of strong acids in DCE. The collection included 62 acids and covered 15 orders of magnitude in acidity, ranging from picric acid, pK$_a$(DCE) = 0.0 (the assigned reference compound), to CF$_3$SO(= NTf)NHTf, pK$_a$(DCE) = –18, where Tf = CF$_3$SO$_2$–.

Based on the values obtained by Kütt et al. [448], Trummal and co-workers [449] applied SMD/B3LYP/6-31G* and SMD/M05-2X/6-31G* calculations with the COSMO-RS continuum solvent model to estimate the acidities in DCE. The results showed very good correlations with the experimental values: r^2 = 0.990 for the M05-2X functional and r^2 = 0.984 for the B3LYP functional.

8.6   OTHER SOLVENTS AND COMMENTARY

Relative and quantitative acidities in many other solvents have been measured. These solvents have included diethyl ether [396], benzene [397], isopropyl alcohol [521,522], n-hexane, heptane [523,524], ENREF [536], *N,N*-dimethylformamide (DMF) [525], methanol [522,526,527], ethanol [522,526], *tert*-butanol [522], cyclohexylamine [528], glacial acetic acid [529], and *N*-methylpyrrolidin-2-one [530].

In each solvent one finds the intrinsic (gas-phase) acidity of the acid changed to a greater or lesser extent by the extrinsic influence of the solvent. The latter influence arises from the extents of solvation of the (usually undissociated) acid HA and the ions A$^-$ and H$^+$ produced by the dissociation. (Of course in some cases, such as the amines, the initial acid HA = R-NH$_3$$^+$ is itself charged and one of its dissociation products is neutral.) In aqueous solution the product ions are normally individually solvated and presumed to be reasonably separated (however, see Schwartz [280]), but in other solvents, especially those with low polarity and low dielectric constants, such separation may not occur or may be incomplete [529]. In such a solution one can picture the acid dissociation as occurring in two steps: first dissociation to form a solvated ion pair, represented as $[A^- \cdot H^+](s)$, followed (possibly) by dissociation of the ion pair to separated ions [280]:

$$K_i \qquad\qquad K_d$$
$$HA(s) \rightleftharpoons [A^- \cdot H^+](s) \rightleftharpoons A^-(s) + H^+(s)$$

Here K_i is an *ionization constant* and K_d is then a *dissociation constant* [280]. In this case one must then take account of the ion pair in determining the overall ionization/dissociation constant.

The case of hydrogen fluoride (HF) in aqueous solution is interesting in this regard. As noted in Section 7.2 and Table 7.1, HF is curiously weak compared to the other halogen acids, in fact being weaker by a factor of 10^{10} compared to HCl. Giguère has explained this weakness in terms of the formation of ion pairs $[F^- \cdot H_3O^+](aq)$ [531]. According to Giguère's analysis, HF behaves as a weak acid because its especially strong ion pairs resist dissociation—freezing point lowering measurements suggest that only 15% of the $[F^- \cdot H_3O^+](aq)$ ion pairs are dissociated even at infinite dilution.

In solvents with very low dielectric constants, such as tetrahydrofuran ($\varepsilon = 7.58$), cyclohexylamine ($\varepsilon = 4.73$), diethyl ether ($\varepsilon = 4.20$), benzene ($\varepsilon = 2.28$), heptane ($\varepsilon = 1.92$), and hexane ($\varepsilon = 1.89$), the second process above (K_d) does not occur to an appreciable extent, and the acidities obtained in these solvents are referred to as *ion-pair acidities* [524]. The concepts and procedures followed in these systems have been well described by Rõõm et al. [524].

How do the acidities determined in different solvents relate to each other? A number of studies have examined this question [22,267,346,416, 448,504,525,527,530,532,533], the most recent being a study by Raamat et al. [433], who looked at strong, neutral Brønsted acids in water, acetonitrile, 1,2-dichloroethane, and the gas phase. These workers found, as might be expected, that the weaker the solvating power of a solvent, the weaker was an acid in that solvent. Less strongly solvating solvents tended to display a greater range of acidities. Exceptions to the observed trends occurred when acids had large, charge-delocalized anions, a feature the authors attribute to the associated larger cavitation energy for these ions. They conclude that the relative acidity changes observed in different solvents can be largely explained in terms of anion charge delocalization.

As a rule the pK_as in different media, since they are all dependent upon the inherent gas-phase acidities of the compounds, tend to show strong correlations, despite some differences in acidity order and other variations that arise from features of individual acids and solvents. Chantooni and Kolthoff [525] compared acidities of substituted phenols in water, DMSO, DMF, and AN, finding good linear, parallel Hammett plots for the pK$_a$s in the three nonaqueous solvents. Other experiments also bear this out. Koppel et al. [416] compared the pK$_a$s of carbon acids in the gas phase and DMSO, and observed linear relationships between the pK$_a$ values in the two solvents for different classes of compounds. For small categories of compounds they found correlation coefficients ranging from 0.888 to 0.999 between the gas-phase and DMSO pK$_a$ values. Seybold and Krey [22] examined the acidities of alcohols and azoles in gas phase, DMSO, and water. They found for the most part that the acidities in these different phases were strongly correlated—r = 0.947–0.991 for the alcohols and r = 0.819–0.988 for the azoles. pK$_a$ values in water and DMSO showed the highest correlations for both sets of compounds.

Recently Himmel et al. have proposed a "unified pH scale for all phases" [534]. This general Brønsted acidity scale is based on the absolute chemical potential of the proton in any medium. They set the standard chemical potential of the proton in the gas phase $\mu_{abs}°(H^+, gas)$ to 0 kJ/mol as a reference value. They report the Gibbs energies of proton solvation $\Delta_{solv}°G(H^+)$ (based on a reference value –1105 ± 8 kJ/mol for water) in most common solvents. Absolute pH values are assigned according to the relation

$$pH_{abs} = \mu_{abs}°(H^+, solv)/(-5.71\ kJ/mol).$$

We note that –1105 kj/mol translates to –264.1 ± 1.9 kcal/mol, so this value aligns most closely with the standard state of 1 atm, but the error bars encompass the 1 M standard state, as explained in Chapter 2.

9 Additional Factors Influencing Acidity and Basicity

In previous chapters the primary emphasis has been placed upon how electronic factors influence the acidities of the compounds examined. For the most part, the pK_as were studied and compared in aqueous solutions at 25°C, although in Chapter 8 attention was also directed at acid-base behavior in other solvents. In this chapter we examine some additional factors that influence acidity and basicity, both in their independent roles and in their influences on the electronic factors responsible for ionization.

9.1 THERMODYNAMICS

Themodynamic considerations of the acid dissociation reaction start with the basic dissociation equilibrium:

$$HA(aq) \rightleftharpoons A^-(aq) + H^+(aq) \tag{9.1}$$

where $K_a = [A^-][H^+]/[HA]$, and by definition $pK_a = -\log K_a$. The Gibbs energy change for this equilibrium emphasizes the relations

$$\Delta G = -RT \ln K_a = -2.303\ RT \log K_a = 2.303\ RT\ pK_a \tag{9.2}$$

and

$$\Delta G = \Delta H - T\Delta S. \tag{9.3}$$

Traditionally attention has been directed toward the change in the dissociation enthalpy, $\Delta H°$, on the assumption that the entropy change $\Delta S°$ for a series of related compounds should remain relatively constant. However, as Calder and Barton [275] and others [535] have noted, in the case of carboxylic acids dissociation $\Delta H°$ values tend to be rather small ($\Delta H° \approx 0 \pm 2$ kcal/mol), and variations in the pK_a values of these compounds tend to reflect mostly variations in the $T\Delta S°$ terms, which often are 5–10 times greater than the $\Delta H°$ terms. In aqueous solution these figures reflect the dissociation reaction

$$R\text{-}COOH(aq) \rightleftharpoons R\text{-}COO^-(aq) + H^+(aq)$$

and the entropy changes arise principally from the ordering of the solvent around the charged ions [536]. For example, the $\Delta H°$ value for the aqueous dissociation of formic acid is -41 cal/mol, while the $T\Delta S°$ value for this dissociation is -3440 cal/mol, and the values for acetic acid are $\Delta H°$ -137 cal/mol and $T\Delta S°$ -6570 cal/mol [275].

This situation is reversed for amine dissociations in aqueous solutions, where the $\Delta H°$ term does in fact dominate the $T\Delta S°$ term [93,331]. Here the relevant dissociation process is

$$R\text{-}NH_3^+(aq) \rightleftharpoons R\text{-}NH_2(aq) + H^+(aq)$$

and a strongly solvated ionic species appears on each side of the equilibrium, largely cancelling the solvent-ordering effect and reducing $\Delta S°$ to a small value. In this case the $\Delta H°$ values are generally much larger than the $T\Delta S°$ terms. For ammonia in water at 25°C, for example, $\Delta H° = 12.5$ kcal/mol and $T\Delta S°$ is just -127 cal/mol, and for aqueous methylamine $\Delta H° = 13.2$ kcal/mol and $T\Delta S°$ is -1350 cal/mol [331].

These are general trends, and can be a helpful guide, but special circumstances can alter these conditions. A further caution, as we shall see in the next section, is that the thermodynamic variables themselves vary with temperature.

9.2   TEMPERATURE EFFECTS ON ACIDITY

The pK$_a$ of a compound can change considerably with temperature, and two general approaches can be employed to understand this variation: a thermodynamic approach and a mechanistic one. The thermodynamic approach to the temperature variation in pK$_a$ is embodied in the *van't Hoff equation*,

$$d(\ln K)/d(1/T) = \Delta H°/R. \tag{9.4a}$$

This can also be written as

$$d(\ln K_a)/dT = \Delta H°/RT^2, \tag{9.4b}$$

or

$$d(pK_a)/dT = \Delta H°/2.303 \, RT^2. \tag{9.4c}$$

It is frequently assumed, without strong justification, that both $\Delta H°$ and $\Delta S°$ are constants and independent of temperature. Accordingly, when the dissociation is exothermic ($\Delta H° < 0$) the pK_a will increase (decreasing acidity) with increasing temperature, and when it is endothermic ($\Delta H° > 0$) the pK_a will decrease (increasing acidity). These variations also are in line with *Le Châtelier's principle*, which holds that higher temperatures should disfavor a process, in this case a dissociation reaction, that releases heat, and favor a reaction that absorbs heat.

D. D. Perrin [537] was among the first researchers to examine the temperature dependence of pK_a in a systematic way, noting that for many dissociations of the form $BH^+ \rightleftharpoons B + H^+$ a temperature rise of 10°C can lead to a decrease of 0.1 to 0.3 pK_a units in the observed pK_a. From the fundamental thermodynamic relation $-\partial\Delta G/\partial T = \Delta S$, one can obtain the expression

$$-d(pK_a)/dT = pK_a/T + \Delta S°/2.303\ RT,$$

indicating that the variation of pK_a with temperature should depend on the starting pK_a of the compound. From typical values for $\Delta S°$ of the dissociation reaction, Perrin showed that for temperatures near 25°C one can approximate the temperature variation of pK_a for many compounds using the simpler expression

$$-d(pK_a)/dT = (pK_a - 0.9)/T \pm 0.004.$$

Perrin presented data for a number of compounds that support the use of this approximation. Notable exceptions include several amine compounds.

Dissociations of the form $AH \rightleftharpoons A^- + H^+$, since they produce two strongly solvated ions from a less strongly solvated neutral species, should lead to greater, more negative $\Delta S°$ values, and will not generally follow a simple approximation such as that found for the $BH^+ \rightleftharpoons B + H^+$ case. One does, however, generally expect less dissociation and a lower pK_a at higher temperatures for such dissociations.

As Calder and Barton note [275], $\Delta H°$ values for acid dissociations can vary significantly with temperature, in some cases even changing sign. Both acetic acid and benzoic acid, e.g., have positive ionization $\Delta H°$ values below room temperature and negative ionization $\Delta H°$ values above room temperature [275]. The *comparative* strengths of these acids can also change considerably over even a modest temperature range—for example, acetic acid is a weaker acid than isopropyl acetic acid below 30°C, but a stronger acid above that temperature [275].

The $\Delta H°$ variations noted above for acetic acid and benzoic acid are in line with the "dome-shaped" pK_a dependence on temperature first noted by Harned and coworkers for acetic acid and related compounds in the 1930s

[538–541], and later reemphasized by Klotz in an examination of protein folding [542]. In these cases the pK$_a$s first increase with temperature, then level off and later decrease as the temperature is raised still further. Klotz attributed these observed dome-shaped temperature variations to inherent changes in the water solvent structure, which result, for example, in a dome-shaped density variation that reaches a maximum near 4°C.

A more direct mechanistic explanation for the dome-shaped temperature dependence and other dome-shaped phenomena pictures two competing molecular processes, one increasing with temperature and the other decreasing with temperature. For the former one can choose the increasing tendency of bond dissociation to occur as the temperature is increased, and for the latter the decreasing hospitality of the aqueous solvent toward ionic species at higher temperatures. Considering the latter effect, one notes that the dielectric constant of water, a measure of its polarity and "friendliness" toward ions, falls from 88 at 0°C to 78.5 at 25°C to 55.3 at 100°C. In the same way the value K$_w$ = [H$_3$O$^+$][OH$^-$] for water falls from 14.943 at 0°C to 13.997 at 25°C to 13.262 at 50°C.

A rather striking visual illustration of the decreasing receptiveness of polar solvents for ionic/polar species at higher temperatures can be found in the thermochromic behavior of the dye rhodamine B [543–545]. In neutral aqueous and alcoholic solutions this dye exists as an equilibrium mixture of two neutral species: a highly-colored, polar zwitterion (Z) and a colorless, nonpolar lactone (L). At lower temperatures the equilibrium in a solvent such as ethanol or n-propanol strongly favors the zwitterion and the solutions are highly colored. However, increasing the temperature of the solution causes a striking loss of color as the Z $\rightleftharpoons$ L equilibrium shifts toward the nonpolar lactone form as the solvent-dye hydrogen bonds that stabilize the zwitterion are increasingly disrupted by thermal motion.

A further illustration of the decreasing hospitality of water and other polar solvents for ionic species at higher temperatures has been presented in a cellular automata model of acid behavior reported by Kier et al. [546]. This model, designed according to heuristic attributes of the acid-water system, shows that increased temperature of the aqueous solvent alone leads to a decrease in the acid dissociation constant in the absence of any changes in the acid itself.

These mechanistic pictures, of course, are entirely compatible with the thermodynamic analysis presented above, and merely substitute a molecular explanation for the more abstract thermodynamic viewpoint.

Because pK$_a$ measurements found in the literature have been made at a variety of temperatures, some commercial and free pK$_a$ models have incorporated schemes for estimating pK$_a$ temperature dependence. The

SPARC model, for example, employs the van't Hoff equation and parameters taken from experimental measurements of temperature variations for this purpose [547].

9.3 STERIC EFFECTS AND HYDROGEN BONDING

Steric effects can influence pK_as by distorting a molecule's structure from planarity or otherwise disrupting the electronic system of the acid. For example, because of the electron-withdrawing power of its *para*-nitro group, 4-nitrophenol has a pK_a of 7.16, well below that of 9.99 for phenol. However, when the nitro group is forced out of the molecular plane by adjacent methyl groups, as in 3,5-dimethyl-4-nitrophenol, the pK_a rises to 8.25 [173].

4-nitrophenol   3,5-dimethyl-4-nitrophenol

Alternatively, structural influences, such as the presence of a rigid ring, can impose constraints on a system. We have already noted the example of Meldrum's acid in Chapter 7, where the rigid ring structure of this carbon acid leads to a pK_a of 7.32 compared to the pK_a of 15.87 of its open-structured analog dimethyl malonate. A similar comparison holds for the cyclic dimedone ($pK_a = 11.2$) and its open analog 2,4-pentadione ($pK_a = 13.3$).

In other cases internal hydrogen bonding can influence the observed pK_a. A classic example is the comparison of maleic acid, with a *cis*-arrangement leading to internal hydrogen bonding, and fumaric acid with a *trans* relationship and no internal hydrogen bonding.

Maleic acid   Fumaric acid

Maleic acid has pK_as of 1.92 and 6.32, the former representing dissociation of the external OH group and the latter dissociation from the H-bonded OH of the anion. The pK_as of fumaric acid are 3.02 and 4.38.

As in the initial dissociation of maleic acid, internal hydrogen bonding can stabilize a dissociation product and enhance the dissociation. Middleton and Lindsey [548], for example, attributed the especially strong acidities of several perfluoro alcohols to stabilization of the anions of these compounds through hydrogen bonds. Perfluoropinacol has a pK$_a$ of 5.95, presumably because of the favorable hydrogen bonding in its first dissociation product anion, shown below.

$$
\begin{array}{c}
\quad\; CF_3 \quad\; CF_3 \\
\quad\;\; | \qquad\;\; | \\
F_3C - C - C - CF_3 \\
\quad\; | \qquad\;\; | \\
\quad\; O \qquad\; O^{\ominus} \\
\quad\;\; \backslash \quad\;.\cdot{}' \\
\qquad\;\; H
\end{array}
$$

9.4   ISOTOPE EFFECTS

We have already seen in Section 8.1 that deuterated acids tend to exhibit pK$_a$s about 0.5–0.6 pK$_a$ units higher than their protonated counterparts. This direct participation in the bond that is formed or, as in the present case, broken in the reaction is called a *primary isotope effect*. When an isotope influences the reaction but is not a direct participant in the affected bond the result is termed a *secondary isotope effect*.

The most obvious change that occurs when one isotope of an atom is substituted for another is the change in mass, since usually the electronic environment remains essentially the same. And clearly the greatest relative perturbation in mass—100%—takes place when deuterium is substituted for ordinary hydrogen (proteum). Although there are small differences in the lengths of C-D and C-H bonds, the greatest changes occur in the vibrational frequency and *zero-point energy* (ZPE). The vibrational frequency of a C-D bond is about 0.71 that of a C-H bond, and the lower ZPE means that more energy is required to break the bond, which should inhibit its dissociation reaction. This can decrease the rate of dissociation by a factor of 2–6, with resulting smaller changes in the pK$_a$. Changes in the carbon isotope from ^{12}C to ^{13}C represents a much smaller perturbation, and normally just a few percent decrease in the reaction rate.

Secondary isotope effects are more subtle, but can be highly informative regarding organic and enzymatic reaction mechanisms [549]. They are described as "α" or "β" effects according to whether the isotope is separated by one or two bonds from the center of reaction. A good example of a β-deuterium secondary isotope effect is found in the pK$_a$s of HCOOH and

DCOOH. In the early 1960s it became apparent that the pK_a of formic acid was slightly lower than that of its deuterated counterpart DCOOH, although experimental uncertainties prevented an exact determination of the difference. This prompted Bell and Miller to undertake a careful experimental determination, which found pK_a(HCOOH) = 3.737 and pK_a(DCOOH) = 3.772, with ΔpK_a = 0.035 ± 0.002 [550]. This was in excellent agreement with a theoretical estimate of 0.037 by Bell and Crooks [551], who attributed the major effect to the zero-point energy.

C.L. Perrin and his students have measured β-deuterium secondary isotope effects on amine basicities and found that deuteration increases the basicities of the compounds investigated [552]. The effect is attributed to a lowering of the ZPE of a C-H bond adjacent to an amine nitrogen. The Perrin group has also applied the same techniques to carboxylic acids and phenols [553], and pyridines [554].

10 Conclusions

After the historical pioneering efforts of Louis de Broglie, Erwin Schrödinger, and the other great minds that gave us quantum mechanics, it was Paul Dirac who wrote in 1929 [555]:

> The underlying physical laws necessary for the mathematical theory of a large part of physics and the whole of chemistry are thus completely known, and the difficulty is only that the exact application of these laws leads to equations much too complicated to be soluble. It therefore becomes desirable that approximate practical methods of applying quantum mechanics should be developed, which can lead to an explanation of the main features of complex atomic systems without too much computation.

We have seen in Chapters 2 and 3 that the modern power of quantum chemistry and the ingenuity of researchers who have developed methods for simulating solvation have allowed for highly accurate relative and absolute pK_a calculations for small molecules. In Chapters 4–9 we have seen how quantitative structure-acidity methods can successfully be used to estimate pK_as on a large number of systems spanning a range of molecular structures and solvents.

In our time, computers have become key scientific instruments that continue to become faster, and cheaper, making enhanced computational capabilities more accessible to research groups with each passing year. We can look forward to a future where highly accurate calculations will be made using explicit solvent models with a high degree of configurational flexibility.

References

[1] B. Saxton and H.F. Meier, *The ionization constants of benzoic acid and of the three monochlorobenzoic acids, at 25°, from conductance measurements*, J. Am. Chem. Soc. 56 (1934), pp. 1918–1921.

[2] E.C. Sherer, G.M. Turner, and G.C. Shields, *Investigation of the potential-energy surface for the first step in the alkaline-hydrolysis of methyl acetate*, Int. J. Quantum Chem. (1995), pp. 83–93.

[3] G.M. Turner, E.C. Sherer, and G.C. Shields, *A computationally efficient procedure for modeling the first step in the alkaline-hydrolysis of esters*, Int. J. Quantum Chem. (1995), pp. 103–112.

[4] E.C. Sherer, G.M. Turner, T.N. Lively, D.W. Landry, and G.C. Shields, *A semiempirical transition state structure for the first step in the alkaline hydrolysis of cocaine. Comparison between the transition state structure, the phosphonate monoester transition state analog, and a newly designed thiophosphonate transition state analog*, J. Mol. Modeling 24 (1996), pp. 62–69.

[5] E.C. Sherer, G. Yang, G.M. Turner, G.C. Shields, and D.W. Landry, *Comparison of experimental and theoretical structures of a transition state analogue used for the induction of anti-cocaine catalytic antibodies*, J. Phys. Chem. A 101 (1997), pp. 8526–8529.

[6] C.G. Zhan, S.X. Deng, J.G. Skiba, B.A. Hayes, S.M. Tschampel, G.C. Shields, and D.W. Landry, *First-principle studies of intermolecular and intramolecular catalysis of protonated cocaine*, J. Comp. Chem. 26 (2005), pp. 980–986.

[7] A.M. Toth, M.D. Liptak, D.L. Phillips, and G.C. Shields, *Accurate relative pK_a calculations for carboxylic acids using complete basis set and Gaussian-n models combined with continuum solvation methods*, J. Chem. Phys. 114 (2001), pp. 4595–4606.

[8] M.D. Liptak and G.C. Shields, *Accurate pK_a calculations for carboxylic acids using complete basis set and Gaussian-n models combined with CPCM continuum solvation methods*, J. Am. Chem. Soc. 123 (2001), pp. 7314–7319.

[9] M. D. Liptak and G.C. Shields, *Experimentation with different thermodynamic cycles used for pK_a calculations on carboxylic acids using complete basis set and Gaussian-n models combined with CPCM continuum solvation methods*, Int. J. Quantum Chem. 85 (2001), pp. 727–741.

[10] E.K. Pokon, M.D. Liptak, S. Feldgus, and G.C. Shields, *Comparison of CBS-QB3, CBS-APNO, and G3 predictions of gas phase deprotonation data*, J. Phys. Chem. A 105 (2001), pp. 10483–10487.

[11] M.D. Liptak, K.C. Gross, P.G. Seybold, S. Feldgus, and G.C. Shields, *Absolute pK_a determinations for substituted phenols*, J. Am. Chem. Soc. 124 (2002), pp. 6421–6427.

[12] M.D. Liptak and G.C. Shields, *Comparison of density functional theory predictions of gas-phase deprotonation data*, Int. J. Quantum Chem. 105 (2005), pp. 580–587.

[13] F.C. Pickard, D.R. Griffith, S.J. Ferrara, M.D. Liptak, K.N. Kirschner, and G.C. Shields, *CCSD(T), W1, and other model chemistry predictions for gas-phase deprotonation reactions*, Int. J. Quantum Chem. 106 (2006), pp. 3122–3128.

[14] J. Ho and M.L. Coote, *pK$_a$ calculation of some biologically important carbon acids—An assessment of contemporary theoretical procedures*, J. Chem. Theory Comput. 5 (2009), pp. 295–306.

[15] K. Alongi and G. Shields, *Theoretical calculations of acid dissociation constants: A review article*, Annu. Rep. Comput. Chem. 6 (2010), pp. 113–138.

[16] J. Ho and M.L. Coote, *First-principles prediction of acidities in the gas and solution phase*, Wiley Interdiscip. Rev.-Comput. Mol. Sci. 1 (2011), pp. 649–660.

[17] K.C. Gross and P.G. Seybold, *Substituent effects on the physical properties and pK$_a$ of aniline*, Int. J. Quantum Chem. 80 (2000), pp. 1107–1115.

[18] K.C. Gross and P.G. Seybold, *Substituent effects on the physical properties and pK$_a$ of phenol*, Int. J. Quantum Chem. 85 (2001), pp. 569–579.

[19] P.G. Seybold, *Analysis of the pK$_a$s of aliphatic amines using quantum chemical descriptors*, Int. J. Quantum Chem. 108 (2008), pp. 2849–2855.

[20] C.A. Hollingsworth, P.G. Seybold, and C.M. Hadad, *Substituent effects on the electronic structure and pK$_a$ of benzoic acid*, Int. J. Quantum Chem. 90 (2002), pp. 1396–1403.

[21] W.C. Kreye and P.G. Seybold, *Correlations between quantum chemical indices and the pK$_a$s of a diverse set of organic phenols*, Int. J. Quantum Chem. 109 (2009), pp. 3679–3684.

[22] P.G. Seybold and W.C. Kreye, *Theoretical estimation of the acidities of alcohols and azoles in gas phase, DMSO, and water*, Int. J. Quantum Chem. 112 (2012), pp. 3769–3776.

[23] M. Simons, A. Topper, B. Sutherland, and P.G. Seybold, *A class project combining organic chemistry, quantum chemistry, and statistics*, Annu. Rev. Comput. Chem. 7 (2011), pp. 237–249.

[24] K.C. Gross, P.G. Seybold, and C.M. Hadad, *Comparison of different atomic charge schemes for predicting pK$_a$ variations in substituted anilines and phenols*, Int. J. Quantum Chem. 90 (2002), pp. 445–458.

[25] K.C. Gross, C.M. Hadad, and P.G. Seybold, *Charge competition in halogenated hydrocarbons*, Int. J. Quantum Chem. 112 (2012), pp. 219–229.

[26] K.C. Gross, P.G. Seybold, Z. Peralta-Inga, J.S. Murray, and P. Politzer, *Comparison of quantum chemical parameters and Hammett constants in correlating pK$_a$ values of substituted anilines*, J. Org. Chem. 66 (2001), pp. 6919–6925.

[27] Y. Ma, K.C. Gross, C.A. Hollingsworth, P.G. Seybold, and J.S. Murray, *Relationships between aqueous acidities and computed surface-electrostatic potentials and local ionization energies of substituted phenols and benzoic acids*, J. Mol. Modeling 10 (2004).

[28] R.D. Rossi, *What does the acid ionization constant tell you? An organic chemistry student guide*, J. Chem. Educ. 90 (2013), pp. 183–190.

[29] A.C. Lee and G.M. Crippen, *Predicting pK$_a$*, J. Chem. Inf. Model. 49 (2009), pp. 2013–2033.

[30] D.A. McQuarrie and J.D. Simon, *Physical Chemistry: A Molecular Approach*, University Science Books, 1997, p. 1360.

[31] S.J. Hawkes, *pKw is almost never 14.0*, J. Chem. Educ. 72:9 (1995), pp. 799–801.

[32] C.E. Moore, B. Jaselskis, and J. Florian, *Historical development of the hydrogen ion concept*, J. Chem. Educ. 87 (2010), pp. 922–923.

[33] T.P. Silverstein, *The solvated proton is NOT H$_3$O$^+$!* J. Chem. Educ. 88 (2011), pp. 875–875.

[34] G.J. Tawa, I.A. Topol, S.K. Burt, R.A. Caldwell, and A.A. Rashin, *Calculation of the aqueous free energy of the proton*, J. Chem. Phys. 109 (1998), pp. 4852–4863.

[35] Y.E. Zevatskii and D.V. Samoilov, *Modern methods for estimation of ionization constants of organic compounds in solution*, Russ. J. Org. Chem. 47 (2011), pp. 1445–1467.

[36] P.G. Seybold, *Quantum chemical-QSPR estimation of the acidities and basicities of organic compounds*, Adv. Quantum Chem. 64 (2012), pp. 83–104.

[37] J. Ho and M.L. Coote, *A universal approach for continuum solvent pK$_a$ calculations: Are we there yet?* Theor. Chem. Acc. 125 (2010), pp. 3–21.

[38] T.N. Brown and N. Mora-Diez, *Computational determination of aqueous pK$_a$ values of protonated benzimidazoles (part 1)*, J. Phys. Chem. B 110 (2006), pp. 9270–9279.

[39] C.P. Kelly, C.J. Cramer, and D.G. Truhlar, *Adding explicit solvent molecules to continuum solvent calculations for the calculation of aqueous acid dissociation constants*, J. Phys. Chem. A 110 (2006), pp. 2493–2499.

[40] J.R. Pliego, *Thermodynamic cycles and the calculation of pK$_a$*, Chem. Phys. Lett. 367 (2003), pp. 145–149.

[41] V.S. Bryantsev, M.S. Diallo, and W.A. Goddard, *Calculation of solvation free energies of charged solutes using mixed cluster/continuum models*, J. Phys. Chem. B 112 (2008), pp. 9709–9719.

[42] M.D. Tissandier, K.A. Cowen, W.Y. Feng, E. Gundlach, M.H. Cohen, A.D. Earhart, J.V. Coe, and T.R. Tuttle, *The proton's absolute aqueous enthalpy and Gibbs free energy of solvation from cluster-ion solvation data*, J. Phys. Chem. A 102 (1998), pp. 7787–7794.

[43] D.M. Camaioni and C.A. Schwerdtfeger, *Comment on "Accurate experimental values for the free energies of hydration of H+, OH–, and H^{3}O$^+$,"* J. Phys. Chem. A 109 (2005), pp. 10795–10797.

[44] Y. Takano and K.N. Houk, *Benchmarking the conductor-like polarizable continuum model (CPCM) for aqueous solvation free energies of neutral and ionic organic molecules*, J. Chem. Theory Comput. 1 (2005), pp. 70–77.

[45] J.R. Pliego and J.M. Riveros, *Theoretical calculation of pK$_a$ using the cluster-continuum model*, J. Phys. Chem. A 106 (2002), pp. 7434–7439.

[46] D.M. McQuarrie, *Statistical Mechanics*, New York, Harper and Row, 1970, p. 86.

[47] W.L. Jorgensen, J.M. Briggs, and J. Gao, *A priori calculations of pK$_a$s for organic compunds in water—The pK$_a$ of ethane*, J. Am. Chem. Soc. 109 (1987), pp. 6857–6858.

[48] W.L. Jorgensen and J.M. Briggs, *A priori pK$_a$ calculations and the hydrations of organic-anions*, J. Am. Chem. Soc. 111 (1989), pp. 4190–4197.

[49] C. Lim, D. Bashford, and M. Karplus, *Absolute pK$_a$ calculations with continuum dielectric methods*, J. Phys. Chem. 95 (1991), pp. 5610–5620.

[50] G. Schüürmann, M. Cossi, V. Barone, and J. Tomasi, *Prediction of the pK$_a$ of carboxylic acids using the ab initio continuum-solvation model PCM-UAHF*, J. Phys. Chem. A 102 (1998), pp. 6706–6712.

[51] C.G. Zhan and D.A. Dixon, *Absolute hydration free energy of the proton from first-principles electronic structure calculations*, J. Phys. Chem. A 105 (2001), pp. 11534–11540.

[52] C.J. Cramer, *Essentials of Computational Chemistry: Theories and Models*, 2nd ed., Wiley, Chichester, England, 2004, p. 579.

[53] M.W. Palascak and G.C. Shields, *Accurate experimental values for the free energies of hydration of H+, OH–, and H$_3$O$^+$*, J. Phys. Chem. A 108 (2004), pp. 3692–3694.

[54] C.P. Kelly, C.J. Cramer, and D.G. Truhlar, *SM6: A density functional theory continuum solvation model for calculating aqueous solvation free energies of neutrals, ions, and solute-water clusters*, J. Chem. Theory Comput. 1 (2005), pp. 1133–1152.

[55] C.P. Kelly, C.J. Cramer, and D.G. Truhlar, *Aqueous solvation free energies of ions and ion-water clusters based on an accurate value for the absolute aqueous solvation free energy of the proton*, J. Phys. Chem. B 110 (2006), pp. 16066–16081.

[56] W.A. Donald and E.R. Williams, *An improved cluster pair correlation method for obtaining the absolute proton hydration energy and enthalpy evaluated with an expanded data set*, J. Phys. Chem. B 114 (2010), pp. 13189–13200.

[57] A.V. Marenich, C.J. Cramer, and D.G. Truhlar, *Universal solvation model based on solute electron density and on a continuum model of the solvent defined by the bulk dielectric constant and atomic surface tensions*, J. Phys. Chem. B 113 (2009), pp. 6378–6396.

[58] J.E. Bartmess, *Negative ion energetics data*, http://webbook.nist.gov (accessed January 20, 2006).

[59] J.W. Ochterski, G.A. Petersson, and J.A. Montgomery, *A complete basis set model chemistry. 5. Extensions to six or more heavy atoms*, J. Chem. Phys. 104 (1996), pp. 2598–2619.

[60] J.A. Montgomery, J.W. Ochterski, and G.A. Petersson, *A complete basis-set model chemistry. 4. An improved atomic pair natural orbital method*, J. Chem. Phys. 10 (1994), pp. 5900–5909.

[61] G.D. Purvis and R.J. Bartlett, *A full coupled-cluster singles and doubles model—The inclusion of disconnected triples*, J. Chem. Phys. 76 (1982), pp. 1910–1918.

[62] J.D. Watts, J. Gauss, and R.J. Bartlett, *Coupled-cluster methods with non-iterative triple excitations for restricted open-shell Hartree-Fock and other general single determinent reference functions—Energies and analytical gradients*, J. Chem. Phys. 98 (1993), pp. 8718–8733.

[63] Y.S. Lee, S.A. Kucharski, and R.J. Bartlett, *A coupled cluster approach with triple excitations*, J. Chem. Phys. 81 (1984), pp. 5906–5912.

[64] J.D. Watts and R.J. Bartlett, *The inclusion of connected triple excitations in the equation-of-motion coupled-cluster method*, J. Chem. Phys. 101 (1994), pp. 3073–3078.

[65] T.H. Dunning, *Gaussian-basis sets for use in correlated molecular calculations. 1. The atoms boron through neon and hydrogen*, J. Chem. Phys. 90 (1989), pp. 1007–1023.

[66] R. Krishnan and J.A. Pople, *Approximate 4th-order perturbation theory of electron correlation energy*, Int. J. Quantum Chem. 14 (1978), pp. 91–100.

[67] T. Helgaker, W. Klopper, H. Koch, and J. Noga, *Basis-set convergence of correlated calculations on water*, J. Chem. Phys. 106 (1997), pp. 9639–9646.

[68] A. Halkier, T. Helgaker, P. Jorgensen, W. Klopper, H. Koch, J. Olsen, and A.K. Wilson, *Basis-set convergence in correlated calculations on Ne, N-2, and H_2O*, Chem. Phys. Lett. 286 (1998), pp. 243–252.

[69] K.L. Bak, P. Jorgensen, J. Olsen, T. Helgaker, and W. Klopper, *Accuracy of atomization energies and reaction enthalpies in standard and extrapolated electronic wave function/basis set calculations*, J. Chem. Phys. 112 (2000), pp. 9229–9242.

[70] L.A. Curtiss, K. Raghavachari, P.C. Redfern, V. Rassolov, and J.A. Pople, *Gaussian-3 (G3) theory for molecules containing first and second-row atoms*, J. Chem. Phys. 109 (1998), pp. 7764–7776.

[71] J.M.L. Martin and G. de Oliveira, *Towards standard methods for benchmark quality ab initio thermochemistry—W1 and W2 theory*, J. Chem. Phys. 111 (1999), pp. 1843–1856.

[72] F.C. Pickard, E.K. Pokon, M.D. Liptak, and G.C. Shields, *Comparison of CBS-QB3, CBS-APNO, G2, and G3 thermochemical predictions with experiment for formation of ionic clusters of hydronium and hydroxide ions complexed with water*, J. Chem. Phys. 122 (2005).

[73] F.C. Pickard, M.E. Dunn, and G.C. Shields, *Comparison of model chemistry and density functional theory thermochemical predictions with experiment for formation of ionic clusters of the ammonium cation complexed with water and ammonia; atmospheric implications*, J. Phys. Chem. A 109 (2005), pp. 4905–4910.

[74] A.J. Cunningham, J.D. Payzant, and P. Kebarle, *A Kinetic study of the proton hydrate H+(H_2O)n equilibria in the gas phase*, J. Am. Chem. Soc. 94 (1972), pp. 7627–7632.

[75] M. Meot-Ner (Mautner) and L.W. Sieck, *Relative acidities of water and methanol and the stabilities of the dimer anions*, J. Phys. Chem. 90 (1986), pp. 6687–6690.

[76] P. Kebarle, *Gas phase ion thermochemistry based on ion-equilibria. From the ionosphere to the reactive centers of enzymes*, Int. J. Mass Spectrom. 200 (2000), pp. 313–330.

[77] G. da Silva, E.M. Kennedy, and B.Z. Dlugogorski, *Ab initio procedure for aqueous-phase pK_a calculation: The acidity of nitrous acid*, J. Phys. Chem. A 110 (2006), pp. 11371–11376.

[78] K.M. Ervin, J. Ho, and W.C. Lineberger, *Ultraviolet photoelectron spectrum of NO_2^-*, J. Phys. Chem. 92:19 (1988), pp. 5405–5412.

[79] Y. Fu, L. Liu, R.-Q. Li, R. Liu, and Q.-X. Guo, *First-principle predictions of absolute pK$_a$'s of organic acids in dimethyl sulfoxide solution*, J. Am. Chem. Soc. 126 (2004), pp. 814–822.

[80] K. Range, D. Riccardi, Q. Cui, M. Elstner, and D. M. York, *Benchmark calculations of proton affinities and gas-phase basicities of molecules important in the study of biological phosphoryl transfer*, Phys. Chem. Chem. Phys. 7 (2005), pp. 3070–3079.

[81] V.S. Bryantsev, M.S. Diallo, and W.A. Goddard, III. *pK$_a$ calculations of aliphatic amines, diamines, and aminoamides via density functional theory with a Poisson-Boltzmann continuum solvent model*, J. Phys. Chem. A 111 (2007), pp. 4422–4430.

[82] B.Guillot, *A reappraisal of what we have learnt during three decades of computer simulations on water*, J. Mol. Liquids 101 (2002), pp. 219–260.

[83] R.M. Shields, B. Temelso, K.A. Archer, T.E. Morrell, and G.C. Shields, *Accurate predictions of water cluster formation, $(H_2O)_{(n\,=\,2-10)}$*, J. Phys. Chem. A 114 (2010), pp. 11725–11737.

[84] B. Temelso, K.A. Archer, and G.C. Shields, *Benchmark structures and binding energies of small water clusters with anharmonicity corrections*, J. Phys. Chem. A 115 (2011), pp. 12034–12046.

[85] C. Perez, M.T. Muckle, D.P. Zaleski, N.A. Seifert, B. Temelso, G.C. Shields, Z. Kisiel, and B.H. Pate, *Structures of cage, prism, and book isomers of water hexamer from broadband rotational spectroscopy*, Science 336 (2012), pp. 897–901.

[86] C.J. Tsai and K.D. Jordan, *Theoretical-study of the $(H_2O)_6$ cluster*, Chem. Phys. Lett. 213 (1993), 181–188.

[87] K. Kim, K.D. Jordan, and T.S. Zwier, *Low-energy structures and vibrational frequencies of the water hexamer—Comparison with benzene-(H_2O) (6)*, J. Am. Chem. Soc. 116 (1994), pp. 11568–11569.

[88] S.S. Xantheas, C.J. Burnham, and R.J. Harrison, *Development of transferable interaction models for water. II. Accurate energetics of the first few water clusters from first principles*, J. Chem. Phys. 116 (2002), pp. 1493–1499.

[89] D.M. Bates and G.S. Tschumper, *CCSD(T) complete basis set limit relative energies for low-lying water hexamer structures*, J. Phys. Chem. A 113 (2009), pp. 3555–3559.

[90] B. Hartke, *Size-dependent transition from all-surface to interior-molecule structures in pure neutral water clusters*, Phys. Chem. Chem. Phys. 5:2 (2003), pp. 275–284.

[91] A. Lagutschenkov, G.S. Fanourgakis, G. Niedner-Schatteburg, and S.S. Xantheas, *The spectroscopic signature of the "all-surface" to "internally solvated" structural transition in water clusters in the n = 17–21 size regime*, J. Chem. Phys. 122 (2005).

[92] S. Yoo, E. Apra, X.C. Zeng, and S.S. Xantheas, *High-level ab initio electronic structure calculations of water clusters $(H_2O)_{16}$ and $(H_2O)_{17}$: A new global minimum for $(H_2O)_{16}$*, J. Phys. Chem. Lett. 1 (2010), pp. 3122–3127.

[93] K.R. Adam, *New density functional and atoms in molecules method of computing relative pK$_a$ values in solution*, J. Phys. Chem. A 106 (2002), pp. 11963–11972.

[94] J.R. Pliego and J.M. Riveros, *The cluster-continuum model for the calculation of the solvation free energy of ionic species*, J. Phys. Chem. A 105 (2001), pp. 7241–7247.

[95] G.A. Kaminski, *Accurate prediction of absolute acidity constants in water with a polarizable force field: Substituted phenols, methanol, and imidazole*, J. Phys. Chem. B 109 (2005), pp. 5884–5890.

[96] A. Klamt and G. Schüürmann, *COSMO: A new approach to dielectric screening in solvents with explicit expressions for the screening energy and its gradient*, J. Chem. Soc. Perkin Trans. 2 (1993), pp. 799–805.

[97] A. Klamt, V. Jonas, T. Burger, and J.C.W. Lohrenz, *Refinement and parametrization of COSMO-RS*, J. Phys. Chem. A 102 (1998), pp. 5074–5085.

[98] S. Miertus, E. Scrocco, and J. Tomasi, *Electrostatic interaction of a solute with a continuum. A direct utilization of ab initio molecular potentials for the prevision of solvent effects*, Chem. Phys. 55 (1981), pp. 117–129.

[99] V. Barone and M. Cossi, *Quantum calculation of molecular energies and energy gradients in solution by a conductor solvent model*, J. Phys. Chem. A 102 (1998), pp. 1995–2001.

[100] M. Cossi, G. Scalmani, N. Rega, and V. Barone, *New developments in the polarizable continuum model for quantum mechanical and classical calculations on molecules in solution*, J. Chem. Phys. 117 (2002), pp. 43–54.

[101] C.J. Cramer and D.G. Truhlar, *Implicit solvation models: Equilibria, structure, spectra, and dynamics*, Chem. Rev. 99 (1999), pp. 2161–2200.

[102] A.V. Marenich, R.M. Olson, C.P. Kelly, J., C.C., and D.G. Truhlar, *Self-consistent reaction field model for aqueous and nonaqueous solutions based on accurate polarized partial charges*, J. Chem. Theory Comput. 3, (2007), pp. 2011–2033.

[103] C.G. Zhan, J. Bentley, and D.M. Chipman, *Volume polarization in reaction field theory*, J. Chem. Phys. 108 (1998), pp. 177–192.

[104] C.G. Zhan and D.M. Chipman, *Cavity size in reaction field theory*, J. Chem. Phys. 109 (1998), pp. 10543–10558.

[105] C.G. Zhan and D.M. Chipman, *Reaction field effects on nitrogen shielding*, J. Chem. Phys. 110 (1999), pp. 1611–1622.

[106] J. Liu, C.P. Kelley, A.C. Goren, A.V. Marenich, C.J. Cramer, D.G. Truhlar, and C.G. Zhan, *Free energies of solvation with surface, volume, and local electrostatic effects and atomic surface tensions to represent the first solvation shell*, J. Chem. Theory Comput. 6 (2010), pp. 1109–1117.

[107] A. Klamt, F. Eckert, M. Diedenhofen, and M.E. Beck, *First principles calculations of aqueous pK_a values for organic and inorganic acids using COSMO-RS reveal an inconsistency in the slope of the pK_a scale*, J. Phys. Chem. A 107 (2003), pp. 9380–9386.

[108] F. Eckert and A. Klamt, *Accurate prediction of basicity in aqueous solution with COSMO-RS*, J. Comp. Chem. 27 (2006), pp. 11–19.

[109] Z.-K. Jia, D.-M. Du, Z.-Y. Zhou, A.-G. Zhang, and R.-Y. Hou, *Accurate pK_a determinations for some organic acids using an extended cluster method*, Chem. Phys. Lett. 439 (2007), pp. 374–380.

[110] G.C. Shields and K.N. Kirschner, *The limitations of certain density functionals in modeling neutral water clusters*, Synth. React. Inorg. Metal-Org. Nano-Met. Chem. 38 (2008), pp. 32–36.

[111] N. Sadlej-Sosnowska, *Calculation of acidic dissociation constants in water: Solvation free energy terms. Their accuracy and impact*, Theor. Chem. Acc. 118 (2007), pp. 281–293.

[112] J.B. Foresman, T.A. Keith, K.B. Wiberg, J. Snoonian, and M.J. Frisch, *Solvent effects. 5. Influence of cavity shape, truncation of electrostatics, and electron correlation ab initio reaction field calculations*, J. Phys. Chem. 100 (1996), pp. 16098–16104.

[113] J. Andzelm, C. Kolmel, and A. Klamt, *Incorporation of solvent effects into density-functional calculations of molecular energies and geometries*, J. Chem. Phys. 103 (1995), pp. 9312–9320.

[114] M. Cossi, N. Rega, G. Scalmani, and V. Barone, *Energies, structures, and electronic properties of molecules in solution with the C-PCM solvation model*, J. Comp. Chem. 24 (2003), pp. 669–681.

[115] A. Yu, Y. Liu, and Y. Wang, *Ab initio calculations on pK_a values of benzoquinuclidine series in aqueous solvent*, Chem. Phys. Lett. 436 (2007), pp. 276–279.

[116] M. Namazian and S. Halvani, *Calculations of pK_a values of carboxylic acids in aqueous solution using density functional theory*, J. Chem. Thermodyn. 38 (2006), pp. 1495–1502.

[117] D.Q. Gao, P. Svoronos, P.K. Wong, D. Maddalena, J. Hwang, and H. Walker, *pK_a of acetate in water: A computational study*, J. Phys. Chem. A 109 (2005), pp. 10776–10785.

[118] C.C. Chambers, G.D. Hawkins, C.J. Cramer, and D.G. Truhlar, *Model for aqueous solvation based on class IV atomic charges and first solvation shell effects*, J. Phys. Chem. 100 (1996), pp. 16385–16398.

[119] H.A. De Abreu, W.B. De Almeida, and H.A. Duarte, *pK_a calculation of poliprotic acid: Histamine*, Chem. Phys. Lett. 383 (2004), pp. 47–52.

[120] N.A. Caballero, F.J. Melendez, C. Munoz-Cara, and A. Nino, *Theoretical prediction of relative and absolute pK_a values of aminopyridines*, Biophys. Chem. 124 (2006), pp. 155–160.

[121] J.A. Pople, M. Head-Gordon, D.J. Fox, K. Raghavachari, and L.A. Curtiss, *Gaussian-1 theory: A general procedure for prediction of molecular energies*, J. Chem. Phys. 90 (1989), pp. 5622–5629.

[122] L.A. Curtiss, K. Raghavachari, G.W. Trucks, and J.A. Pople, *Gaussian-2 theory for molecular-energies of 1st-row and 2nd-row compounds*, J. Chem. Phys. 94 (1991), pp. 7221–7230.

[123] M.A.K. Liton, M.I. Ali, and M.T. Hossain, *Accurate pK_a calculations for trimethylaminium ion with a variety of basis sets and methods combined with CPCM continuum solvation methods*, Comput. Theor. Chem. 999 (2012), pp. 1–6.

[124] C.C.R. Sutton, G.V. Franks, and G. da Silva, *First principles pK_a calculations on carboxylic acids using the SMD solvation model: Effect of thermodynamic cycle, model chemistry, and explicit solvent molecules*, J. Phys. Chem. B 116 (2012), pp. 11999–12006.

[125] J.A. Keith and E.A. Carter, *Quantum chemical benchmarking, validation, and prediction of acidity constants for substituted pyridinium ions and pyridinyl radicals*, J. Chem. Theory Comput. 8 (2012), pp. 3187–3206.

[126] J. Crugeiras, A. Rios, and H. Maskill, *DFT and AIM study of the protonation of nitrous acid and the pK$_a$ of nitrous acidium ion*, J. Phys. Chem. A 115 (2011), pp. 12357–12363.

[127] J. Ho, M.L. Coote, M. Franco-Perez, and R. Gomez-Balderas, *First-principles prediction of the pK$_a$s of anti-inflammatory oxicams*, J. Phys. Chem. A 114 (2010), pp. 11992–12003.

[128] S. Rayne and K. Forest, *Theoretical studies on the pK$_a$ values of perfluoro-alkyl carboxylic acids*, J. Mol. Struct.-Theochem. 949 (2010), pp. 60–69.

[129] R. Casasnovas, J. Frau, J. Ortega-Castro, A. Salva, J. Donoso, and F. Munoz, *Absolute and relative pK$_a$ calculations of mono and diprotic pyridines by quantum methods*, J. Mol. Struct.-Theochem. 912 (2009), pp. 5–12.

[130] Y.H. Jang, W.A. Goddard, K.T. Noyes, L.C. Sowers, S. Hwang, and D.S. Chung, *pK$_a$ values of guanine in water: Density functional theory calculations combined with Poisson-Boltzmann continuum-solvation model*, J. Phys. Chem. B 107 (2003), pp. 344–357.

[131] V. Verdolino, R. Cammi, B.H. Munk, and H.B. Schlegel, *Calculation of pK$_a$ values of nucleobases and the guanine oxidation products guanidinohydantoin and spiroiminodihydantoin using density functional theory and a polarizable continuum model*, J. Phys. Chem. B 112 (2008), pp. 16860–16873.

[132] N. Sadlej-Sosnowska, *On the way to physical interpretation of Hammett constants: How substituent active space impacts on acidity and electron distribution in p-substituted benzoic acid molecules*, Pol. J. Chem. 81 (2007), pp. 1123–1134.

[133] M. Krol, M. Wrona, C.S. Page, and P.A. Bates, *Macroscopic pK$_a$ calculations for fluorescein and its derivatives*, J. Chem. Theory Comput. 2 (2006), pp. 1520–1529.

[134] M. Sramko, M. Smiesko, and M. Remko, *Accurate aqueous proton dissociation constants calculations for selected angiotensin-converting enzyme inhibitors*, J. Biomol. Struct. Dyn. 25 (2008), pp. 599–608.

[135] A.M. Amado, S.M. Fiuza, L. de Carvalho, P.J.A. Ribeiro-Claro, *On the effects of changing Gaussian program version and SCRF defining parameters: Isopropylamine as a case study*, Bull. Chem. Soc. Jpn. 85 (2012), pp. 962–975.

[136] A.V. Marenich, W.D. Ding, C.J. Cramer, and D.G. Truhlar, *Resolution of a challenge for solvation modeling: Calculation of dicarboxylic acid dissociation constants using mixed discrete-continuum solvation models*, J. Phys. Chem. Lett. 3 (2012), pp. 1437–1442.

[137] Y. Zhao and D.G. Truhlar, *The M06 suite of density functionals for main group thermochemistry, thermochemical kinetics, noncovalent interactions, excited states, and transition elements: Two new functionals and systematic testing of four M06-class functionals and 12 other functionals*, Theor. Chem. Acc. 120 (2008), pp. 215–241.

[138] Y. Zhao and D.G. Truhlar, *Density functionals with broad applicability in chemistry*, Acc. Chem. Res. 41 (2008), pp. 157–167.

[139] B.J. Lynch, Y. Zhao, and D.G. Truhlar, *Effectiveness of diffuse basis functions for calculating relative energies by density functional theory*, J. Phys. Chem. A 107 (2003), pp. 1384–1388.

[140] A.M. Rebollar-Zepeda and A. Galano, *First principles calculations of pKₐ values of amines in aqueous solution: Application to neurotransmitters*, Int. J. Quantum Chem. 112 (2012), pp. 3449–3460.

[141] A. M. Rebollar-Zepeda, T. Campos-Hernandez, M. Teresa Ramirez-Silva, A. Rojas-Hernandez, and A. Galano, *Searching for computational strategies to accurately predict pKₐs of large phenolic derivatives*, J. Chem. Theory Comput. 7 (2011), pp. 2528–2538.

[142] X.-X. Wang, H. Fu, D.-M. Du, Z.-Y. Zhou, A.-G. Zhang, C.-F. Su, and K.-S. Ma, *The comparison of pKₐ determination between carbonic acid and formic acid and its application to prediction of the hydration numbers*, Chem. Phys. Lett. 460 (2008), pp. 339–342.

[143] A.V. Marenich, C.J. Cramer, and D.G. Truhlar, *Perspective on foundations of solvation modeling: The electrostatic contribution to the free energy of solvation*, J. Chem. Theory Comput. 4 (2008), pp. 877–887.

[144] A.V. Marenich, C.J. Cramer, and D.G. Truhlar, *Generalized Born solvation model SM12*, J. Chem. Theory Comput. 9 (2013), pp. 609–620.

[145] A. Amat, S. Fantacci, F. De Angelis, B. Carlotti, and F. Elisei, *DFT/TDDFT investigation of the stepwise deprotonation in tetracycline: pKₐ assignment and UV-vis spectroscopy*, Theor. Chem. Acc. 131 (2012).

[146] C. Matijssen, G.K. Kinsella, G.W. Watson, and I. Rozas, *Computational study of the proton affinity and basicity of structurally diverse a1-adrenoceptor ligands*, J. Phys. Organic Chem. 25 (2012), pp. 351–360.

[147] A. Amat, F. De Angelis, A. Sgamellotti, and S. Fantacci, *Acid-base chemistry of luteolin and its methyl-ether derivatives: A DFT and ab initio investigation*, Chem. Phys. Lett. 462 (2008), pp. 313–317.

[148] C.L. Perrin, M.A. Fabian, and K.B. Armstrong, *Solvation effect on steric bulk of ionic substituents—Imidazolium vs imidazole*, J. Org. Chem. 59 (1994), pp. 5246–5253.

[149] C.L. Perrin and M.A. Fabian, *Multicomponent NMR titration for simultaneous measurement of relative pKₐs*, Anal. Chem. 68 (1996), pp. 2127–2134.

[150] Y. Zeng, X.L. Chen, D.B. Zhao, H.T. Li, Y.Y. Zhang, and X.M. Xiao, *Estimation of pKₐ values for carboxylic acids, alcohols, phenols and amines using changes in the relative Gibbs free energy*, Fluid Phase Equilibria 313 (2012), pp. 148–155.

[151] R. Gomez-Bombarelli, M. Gonzalez-Perez, M.T. Perez-Prior, E. Calle, and J. Casado, *Computational study of the acid dissociation of esters and lactones. A case study of diketene*, J. Org. Chem. 74 (2009), pp. 4943–4948.

[152] K.K. Govender and I. Cukrowski, *Density functional theory in prediction of four stepwise protonation constants for nitrilotripropanoic acid (NTPA)*, J. Phys. Chem. A 113 (2009), pp. 3639–3647.

[153] O.A. Borg and B. Durbeej, *Relative ground and excited-state pKₐ values of phytochromobilin in the photoactivation of phytochrome: A computational study*, J. Phys. Chem. B 111 (2007), pp. 11554–11565.

[154] O.A. Borg and B. Durbeej, *Which factors determine the acidity of the phytochromobilin chromophore of plant phytochrome?* Phys. Chem. Chem. Phys. 10 (2008), pp. 2528–2537.

[155] J.H. Jensen, H. Li, A.D. Robertson, and P.A. Molina, *Prediction and ratio-nalization of protein pK$_a$ values using QM and QM/MM methods*, J. Phys. Chem. A 109 (2005), pp. 6634–6643.

[156] H. Li, A.W. Hains, J.E. Everts, A.D. Robertson, and J.H. Jensen, *The prediction of protein pK$_a$'s using QM/MM: The pK$_a$ of lysine 55 in Turkey ovomucoid third domain*, J. Phys. Chem. B 106 (2002), pp. 3486–3494.

[157] H. Li, A.D. Robertson, and J.H. Jensen, *The determinants of carboxyl pK$_a$ values in Turkey ovomucoid third domain*, Proteins Struct. Funct. Bioinform. 55 (2004), pp. 689–704.

[158] T. Matsui, T. Baba, K. Kamiya, and Y. Shigeta, *An accurate density functional theory based estimation of pK$_a$ values of polar residues combined with experimental data: From amino acids to minimal proteins*, Phys. Chem. Chem. Phys. 14 (2012), pp. 4181–4187.

[159] I. Sharma and G.A. Kaminski, *Calculating pK$_a$ values for substituted phenols and hydration energies for other compounds with the first-order fuzzy-border continuum solvation model*, J. Comp. Chem. 33 (2012), pp. 2388–2399.

[160] F. Ding, J.M. Smith, and H. Wang, *First-principles calculation of pK$_a$ values for organic acids in nonaqueous solution*, J. Org. Chem. 74 (2009), pp. 2679–2691.

[161] G.I. Almerindo, D.W. Tondo, and J.R. Pliego, *Ionization of organic acids in dimethyl sulfoxide solution: A theoretical ab initio calculation of the pK$_a$ using a new parametrization of the polarizable continuum model*, J. Phys. Chem. A 108 (2004), pp. 166–171.

[162] A. Shokri, A. Abedin, A. Fattahi, and S.R. Kass, *Effect of hydrogen bonds on pK$_a$ values: Importance of networking*, J. Am. Chem. Soc. 134 (2012), pp. 10646–10650.

[163] P.G. Seybold, M. May, and U.A. Bagal, *Molecular structure-property relationships*, J. Chem. Educ. 64 (1987), pp. 575–581.

[164] D.E. Needham, I.-C. Wei, and P.G. Seybold, *Molecular modeling of the physical properties of the alkanes*, J. Am. Chem. Soc. 110 (1988), pp. 4186–4194.

[165] S.D. Nelson and P.G. Seybold, *Molecular structure-property relationships for alkenes*, J. Mol. Graph. Model. 20 (2001), pp. 36–53.

[166] A.R. Katritzky, V.S. Lobanov, and M. Karelson, *QSPR: The correlation and quantitative prediction of chemical and physical properties from structure*, Chem. Soc. Rev. 24 (1995), pp. 279–287.

[167] M. Karelson, V.S. Lobanov, and A.R. Katritzky, *Quantum chemical descriptors in QSAR/QSPR studies*, Chem. Rev. 96 (1996), pp. 1027–1043.

[168] A.R. Katritzky, U. Maran, V.S. Lobanov, and M. Karelson, *Structurally diverse quantitative structure-property relationship correlations of technologically relevant physical properties*, J. Chem. Inf. Comput. Sci. 40 (2000), pp. 1–18.

[169] A.R. Katritzky, M. Kuanar, S. Slavov, C.D. Hall, M. Karelson, L. Kahn, and D.A. Dobchev, *Quantitative correlation of physical and chemical properties with chemical structure: Utility for prediction*, Chem. Rev. 110 (2010), pp. 5714–5789.

[170] D.J. Livingstone, *The characterization of chemical structures using molecular properties. A survey*, J. Chem. Inf. Comput. Sci. 40 (2000), pp. 195–209.

[171] J. Catalán and A. Macias, *Intrinsic acidities of meta- and para-substituted phenols from calculated molecular properties*, J. Chem. Soc. Perkin Trans. 2 (1979), pp. 1632–1636.

[172] S. Zhang, J. Baker, and P. Pulay, *A reliable and efficient first principles-based method for predicting pK_a values. 1. Methodology*, J. Phys. Chem. A 114 (2010), pp. 425–431.

[173] D.D. Perrin, B. Dempsey, and E.P. Sergeant, *pK_a Prediction for Organic Acids and Bases*, Chapman & Hall, New York, 1981.

[174] G.E.P. Box and N.R. Draper, *Empirical Model-Building and Response Surfaces*, Wiley, New York, 1987.

[175] L.B. Kier, P.G. Seybold, and C.-K. Cheng, *Modeling Chemical Systems Using Cellular Automata*, Springer, Dordrecht, The Netherlands, 2005.

[176] D. Young, T. Martin, R. Venkatapathy, and P. Harten, *Are the chemical structures in your QSAR correct?* QSAR Comb. Sci. 27 (2008), pp. 1337–1345.

[177] D. Fourches, E. Muratov, and A. Tropsha, *Trust, but verify: On the importance of chemical structure curation in cheminformation and QSAR modeling research*, J. Chem. Inf. Model. 50 (2010), pp. 1189–1204.

[178] P.G. Seybold, *Explorations of molecular structure-property relationships*, SAR QSAR Environ. Res. 10 (1999), pp. 101–115.

[179] H.C. Gauch, Jr., *Prediction, parsimony, and noise*, Amer. Sci. 81 (1993), pp. 468–478.

[180] M.T.D. Cronin and T.W. Schultz, *Pitfalls in QSAR*, J. Mol. Struct.-Theochem. 622 (2003), pp. 39–51.

[181] J.C. Dearden, M.T.D. Cronin, and K.L.E. Kaiser, *How not to develop a quantitative structure–activity or structure–property relationship (QSAR/QSPR)*, SAR QSAR Environ. Res. 20 (2009), pp. 241–266.

[182] D.D. Perrin, *Prediction of pK_a values*, in *Physical Chemical Properties of Drugs*, S.H. Yalkowsky, A.A. Sinkula, and S.C. Valvani, eds., Marcel Dekker, New York, 1980.

[183] L.P. Hammett, *The effect of structure upon the reactions of organic compounds: Benzene derivatives*, J. Am. Chem. Soc. 59 (1937), pp. 96–103.

[184] L.P. Hammett, *Linear free energy relationships in rate and equilibrium phenomena*, Trans. Faraday Soc. (1938), 156–165.

[185] L.P. Hammett, *Physical Organic Chemistry*, McGraw-Hill, New York, 1940.

[186] R.W. Taft, Jr., *Linear free energy relationships from rates of esterification and hydrolysis of aliphatic and ortho-substituted benzoate esters*, J. Am. Chem. Soc. 74 (1952), pp. 2729–2732.

[187] R.W. Taft, Jr., *Polar and steric substituent constants for aliphatic and o-benzoate groups from rates of esterification and hydrolysis of esters*, J. Am. Chem. Soc. 74 (1952), pp. 3120–3128.

[188] R.W. Taft, *Regarding the inherent dependence of resonance effects of strongly conjugated substituents on electron demand*, J. Am. Chem. Soc. 110 (1988), pp. 1797–1800.

[189] H.K. Hall, Jr.. *Field and inductive effects on the base strengths of amines*, J. Am. Chem. Soc. 78 (1956), pp. 2570–2572.

[190] H.K. Hall, Jr. *Correlation of the base strengths of amines*, J. Am. Chem. Soc. 79 (1957), pp. 5441–5444.

[191] P. Ballinger and F.A. Long, *Acid ionization constants of alcohols. II. Acidities of some substituted methanols and related compounds*, J. Am. Chem. Soc. 82 (1960), pp. 795–798.

[192] C. Hansch, A. Leo, and R.W. Taft, *A survey of Hammett substituent constants and resonance and field phenomena*, Chem. Rev. 91 (1991), pp. 165–195.

[193] R.W. Taft, I.A. Koppel, R.D. Topsom, and F. Anvia, *Acidities of OH compounds, including alcohols, phenol, carboxylic acids, and mineral acids*, J. Am. Chem. Soc. 112 (1990), pp. 2047–2052.

[194] H.H. Jaffé, *Correlation of Hammett's σ-values with electron densities calculated by molecular orbital theory*, J. Chem. Phys. 20 (1952), pp. 279–284.

[195] H.H. Jaffé, *Theoretical considerations concerning Hammett's equation. II. Calculation of σ-values for toluene and naphthalene*, J. Chem. Phys. 20 (1952), pp. 778–780.

[196] H.H. Jaffé, *Theoretical considerations concerning Hammett's equation. III. σ-Values for pyradine and other aza-substituted hydrocarbons*, J. Chem. Phys. 20 (1952), pp. 1554–1555.

[197] R.D. Gilliom, J.-P. Beck, and W.P. Purcell, *An MNDO treatment of sigma values*, J. Comp. Chem. 6 (1985), pp. 437–440.

[198] G. Schüürmann, *Do Hammett constants model electronic properties in QSARs?* Sci. Total Environ. 109/110 (1991), pp. 221–235.

[199] M. Haeberlein, J.S. Murray, T. Brinck, and P. Politzer, *Calculated electrostatic potentials and local surface ionization energies of para-substitituted anilines as measures of substituent effects*, Can. J. Chem. 70 (1992), pp. 2209–2214.

[200] K.H. Kim and Y.C. Martin, *Direct prediction of linear free energy substituent effects from 3D structures using comparative molecular field analysis. 1. Electronic effects of substituted benzoic acids*, J. Org. Chem. 56 (1991), pp. 2723–2729.

[201] Y. Takahata and D.P. Chong, *Estimation of Hammett sigma constants of substituted benzenes through accurate density-functional calculation of core-electron binding energy shifts*, Int. J. Quantum Chem. 103 (2005), pp. 509–515.

[202] L. Rincón and R. Almeida, *Is the Hammett's constant free of steric effects?* J. Phys. Chem. A 116 (2012), pp. 7523–7530.

[203] P. Ertl, *Simple quantum chemical parameters as an alternative to Hammett sigma constants in QSAR studies*, Quant. Struct.-Act. Relat. 16 (1997), pp. 377–382.

[204] J.J. Sullivan, A.D. Jones, and K.K. Tanji, *QSAR treatment of electronic substituent effects using frontier orbital theory and topological parameters*, J. Chem. Inf. Comput. Sci. 40 (2000), pp. 1113–1127.

[205] Y. Simón-Manso, *Linear free-energy relationships and the density functional theory: An analog of the Hammett equation*, J. Phys. Chem. A 109 (2005), pp. 2006–2011.

[206] K.B. Wiberg and P.R. Rablen, *Comparison of different atomic charges derived via different procedures*, J. Comp. Chem. 14 (1993), pp. 1504–1518.

[207] R.F.W. Bader, *Atoms in Molecules: A Quantum Theory*, Clarendon Press, Oxford, 1990.

[208] C.L. Perrin, *Atomic size dependence of Bader electron populations: Significance for questions of resonance stabilization*, J. Am. Chem. Soc. 113 (1991), pp. 2865–2868.

[209] S.L. Dixon and P.C. Jurs, *Estimation of pK_a for organic oxyacids using calculated atomic charges*, J. Comp. Chem. 14 (1993), pp. 1460–1467.

[210] M.J. Citra, *Estimatic the pK_a of phenols, carboxylic acids and alcohols from semi-empirical quantum chemical methods*, Chemosphere 38 (1999), pp. 191–206.

[211] M.J.S. Dewar, E.G. Zoebisch, E.F. Healy, and J.J.P. Stewart, *Development and use of quantum mechanical molecular models. 76. AM1: A new general purpose quantum mechanical molecular model*, J. Am. Chem. Soc. 107 (1985), pp. 3902–3909.

[212] B.G. Tehan, E.J. Lloyd, M.G. Wong, W.R. Pitt, J.G. Montana, D.T. Manallack, and E. Gancia, *Estimation of pK_a using semiempirical molecular orbital methods. Part 1. Application to phenols and carboxylic acids*, Quant. Struct.-Activity Relationships 21 (2002), pp. 457–472.

[213] B.G. Tehan, E.J. Lloyd, M.G. Wong, W.R. Pitt, E. Gancia, and D.T. Manallack, *Estimation of pK_a using semiempirical molecular orbital methods. Part 2. Application to amines, anilines and various nitrogen containing heterocyclic compounds*, Quant. Struct.-Activity Relationships 21 (2002), pp. 473–485.

[214] K. Fukui, T. Yonezawa, and C. Nagata, *Theory of substitution in conjugated molecules*, Bull. Chem. Soc. Jpn. 27 (1954), pp. 423–427.

[215] S. Jelfs, P. Ertl, and P. Selzer, *Estimation of pK_a for druglike compounds using semiempirical and information-based descriptors*, J. Chem. Inf. Model. 47 (2007), pp. 450–459.

[216] F.A. La Porta, R.T. Santiago, T.C. Ramalho, M.P. Freitas, and E.F.F. Da Cunha, *The role of frontier orbitals in acid-base chemistry of organic amines probed by ab initio and chemometric techniques*, Int. J. Quantum Chem. 110 (2010), pp. 2015–2023.

[217] J. Zhang, T. Kleinöder, and J. Gasteiger, *Prediction of pK_a values for aliphatic carboxylic acids and alcohols with empirical atomic charge descriptors*, J. Chem. Inf. Model. 46 (2006), pp. 2256–2266.

[218] B.H. Besler, K.M. Merz, and P.A. Kollman, *Atomic charges derived from semiempirical methods*, J. Comp. Chem. 11 (1990), pp. 431–439.

[219] J. Cioslowski, A new population analysis based on atomic polar tensors, J. Am. Chem. Soc. 111 (1989), pp. 8333–8336.

[220] R.S. Mulliken, *Criteria for the construction of good self-consistent-field molecular orbital wave functions, and the significance of LCAO-MO population analysis*, J. Chem. Phys. 36 (1962), pp. 3428–3439.

[221] A.E. Reed, R.B. Weinstock, and F. Weinhold, *Natural popoulation analysis*, J. Chem. Phys. 83 (1985), pp. 735–746.

[222] P.-O. Löwdin, *On the nonorthogonality problem*, Adv. Quantum Chem. 5 (1970), pp. 185–199.

[223] J. Gasteiger and M. Marsili, *Iterative partial equalization of orbital electronegativity—A rapid access to atomic charges*, Tetrahedron 36 (1980), pp. 3219–3228.

[224] W.J. Mortier, K. Van Genechten, and J. Gasteiger, *Electronegativity equalization: Application and parametrization*, J. Am. Chem. Soc. 107 (1985), pp. 829–835.

[225] R.S. Vařeková, S. Geidl, C.-M. Ionescu, O. Skřehota, M. Kudera, D. Sehnal, T. Bouchal, R. Abagyan, H.J. Huber, and J. Koča, *Predicting pK_a values of substituted phenols from atomic charges: Comparison of different quantum mechanical methods and charge distribution schemes*, J. Chem. Inf. Model. 51 (2011), pp. 1795–1806.

[226] J.A. Pople and D.L. Beveridge, *Approximate Molecular Orbital Theory*, McGraw-Hill, New York, 1970.

[227] G. La Manna, V. Tschinke, and L. Paoloni, *Theoretical correlation of substituent effects on the acidity of benzoic acids in the vapour phase with calculated HOMO eigenvalues*, J. Chem. Soc. Perkin Trans. 2 (1985), pp. 1393–1394.

[228] T. Sotomatsu, Y. Murata, and T. Fujita, *Correlation analysis of substituent effects on the acidity of benzoic acids by the AM1 method*, J. Comp. Chem. 10 (1989), pp. 94–98.

[229] C. Grüber and V. Buss, *Quantum-mechanically calculated properties for the development of quantitative structure-activity relationships (QSARs). pK_a values of phenols and aromatic and aliphatic carboxylic acids*, Chemosphere 19 (1989), pp. 1595–1609.

[230] S. Zhang, J. Baker, and P. Pulay, *A reliable and efficient first principles-based method for predicting pK_a values. 2. Organic acids*, J. Phys. Chem. A 114 (2010), pp. 432–442.

[231] T. Brinck, J.S. Murray, P. Politzer, and R.E. Carter, *A relationship between experimentally determined pK_as and molecular surface ionization energies for some azines and azoles*, J. Org. Chem. 56 (1991), pp. 2934–2936.

[232] T. Sakai, T. Korenaga, N. Washio, Y. Nishio, S. Minami, and T. Ema, *Synthesis of enantiomerically pure (R,R)- and (S,S)-1,2-bis(pentafluorophenyl)ethane-1,2-diamine and evaluation of the pK_a value by ab initio calculations*, Bull. Chem. Soc. Jpn. 77 (2004), pp. 1001–1008.

[233] J. Han, R.L. Deming, and F.-M. Tao, *Theoretical study of molecular structures and properties of the complete series of chlorophenols*, J. Phys. Chem. A 108 (2004), pp. 7736–7743.

[234] J. Han, H. Lee, and F.-M. Tao, *Molecular structures and properties of the complete series of bromophenols: Density functional theory calculations*, J. Phys. Chem. A 109 (2005), pp. 5186–5192.

[235] J. Han and F.-M. Tao, *Correlations and predictions of pK_a values of fluorophenols and bromophenols using hydrogen-bonded complexes with ammonia*, J. Phys. Chem. A 110 (2006), pp. 257–263.

[236] J. Han, R.L. Deming, and F.-M. Tao, *Theoretical study of hydrogen-bonded complexes of chlorophenols with water or ammonia: Correlations and predictions of pK_a values*, J. Phys. Chem. A 109 (2005), pp. 1159–1167.

[237] L. Tao, J. Han, and F.-M. Tao, *Correlations and predictions of carboxylic acid pK_a values using intermolecular structure and properties of hydrogen-bonded complexes*, J. Phys. Chem. A 112 (2008), pp. 775–782.

[238] R. Parthaserathi, J. Padmanabhan, M. Elango, K. Chitra, V. Subramannian, and P.K. Chattaraj, *pK_a prediction using group philicity*, J. Phys. Chem. A 110 (2006), pp. 6540–6544.

[239] P.K. Chattaraj, A. Chakraborty, and S. Giri, *Net electrophilicity*, J. Phys. Chem. A 113 (2009), pp. 10068–10074.

[240] R.G. Parr, L.V. Szentpály, and S. Liu, *Electrophilicity index*, J. Am. Chem. Soc. 121 (1999), pp. 1922–1924.

[241] K. Gupta, S. Giri, and P.K. Chattaraj, *Acidity of meta- and para-substituted aromatic acids: A conceptual DFT study*, New J. Chem. 32 (2008), p. 1945.

[242] R.D. Cramer, D.E. Patterson, and J.D. Bunce, *Comparative molecular-field analysis (COMFA). 1. Effect of shape on binding of steroids to carrier proteins*, J. Am. Chem. Soc. 110 (1988), pp. 5959–5967.

[243] R. Gargallo, C.A. Sotriffer, K.R. Liedl, and B.M. Rode, *Application of multivariate data analysis methods to comparative molecular field analysis (CoMFA) data: Proton affinities and pK_a prediction for nucleic acids components*, J. Comput.-Aided Mol. Des. 13 (1999), pp. 611–623.

[244] L. Xing and R.C. Glen, *Novel methods for the prediction of logP, pK_a, and logD*, J. Chem. Inf. Comput. Sci. 42 (2002), pp. 796–805.

[245] L. Xing, R.C. Glen, and R.D. Clark, *Predicting pK_a by molecular tree structured fingerprints and PLS*, J. Chem. Inf. Comput. Sci. 43 (2003), pp. 870–879.

[246] W. Lindberg, J.-A. Persson, and S. Wold, *Partial least squares method for spectrofluorimetric analysis of mixtures of humic acid and ligninsulfonate*, Anal. Chem. 55 (1983), pp. 643–648.

[247] A.C. Lee, J.-Y. Yu, and G.M. Crippen, *pK_a prediction of monoprotic small molecules the SMARTS way*, J. Chem. Inf. Model. 48 (2008), pp. 2042–2053.

[248] F. Milletti, L. Storchi, G. Sforna, and G. Cruciani, *New and original pK_a prediction method using grid molecular interaction fields*, J. Chem. Inf. Model. 47 (2007), pp. 2172–2181.

[249] P.J. Goodford, *A computational procedure for determining energetically favorable binding sites on biologically important macromolecules*, J. Med. Chem. 28 (1985), pp. 849–857.

[250] F. Luan, W. Ma, H. Zhang, X. Zhang, M. Liu, Z. Hu, and B. Fan, *Prediction of pK_a for neutral and basic drugs based on radial basis function neural networks and the heuristic method*, Pharm. Res. 22 (2005), pp. 1454–1460.

[251] A. Habibi-Yangjeh, M. Danandeh-Jenagharad, and M. Nooshyar, *Prediction acidity constant of various benzoic acids and phenols in water using linear and nonlinear QSPR models*, Bull. Korean Chem. Soc. 26 (2005), pp. 2007–2016.

[252] A. Habibi-Yangjeh, M. Danandeh-Jenagharad, and M. Nooshyar, *Application of artificial neural networks for predicting the aqueous acidity of various phenols using QSAR*, J. Mol. Modeling 12 (2006), pp. 338–347.

[253] A. Habibi-Yangjeh and M. Danandeh-Jenagharad, *Prediction of acidity constant for substituted acetic acids in water using artificial neural networks*, Indian J. Chem., Sect. B: Org. Chem. Incl. Med. Chem. 46 (2007), pp. 478–487.

[254] A. Habibi-Yangjeh, E. Pourbasheer, and M. Danandeh-Jenagharad, *Application of principal component-genetic algorithm-artificial neural network for prediction acidity constant of various nitrogen-containing compounds in water*, Monatsh. Chem. 140 (2009), pp. 15–27.

[255] S.H. Hilal, S.W. Karickhoff, and L.A. Carrreira, *A rigorous test for SPARC's chemical reactivity models: Estimation of more than 4300 ionization pK_as*, Quant. Struct.-Activity Relationships 14 (1995), pp. 348–355.

[256] C. Liao and M.C. Nicklaus, *Comparison of nine programs predicting pK_a values of pharmaceutical substances*, J. Chem. Inf. Model. 49 (2009), pp. 2801–2812.

[257] H. Yu, R. Kühne, R.-U. Ebert, and G. Schüürmann, *Comparative analysis of QSAR models for predicting pK_a of organic oxygen acids and nitrogen bases from molecular structure*, J. Chem. Inf. Model. 50 (2010), pp. 1949–1960.

[258] J. Manchester, G. Walkup, and Z. You, *Evaluation of pK_a estimation methods on 211 druglike compound*, J. Chem. Inf. Model. 50 (2010), pp. 565–571.

[259] J.C. Shelley, D. Calkins, and A.P. Sullivan, *Comments on the article "Evaluation of pK_a estimation methods on 211 druglike compounds,"* J. Chem. Inf. Model. 51 (2011), pp. 102–104.

[260] J.C. Dearden, M.T.D. Cronin, and D.C. Lappin, *A comparison of commercially available software for the prediction of pK_a*, J. Pharmacy Pharmacol. (2007), Supplement 1, A7.

[261] J.C. Dearden, P. Rotureau, and G. Fayet, *QSPR prediction of physicochemical properties for REACH*, SAR QSAR Environ. Res. (2013), pp. 1–39.

[262] D.R. Lide, *CRC Handbook of Chemistry and Physics*, Vol. 90, CRC Press, Boca Raton, FL, 2009–2010.

[263] J. Murto, *Nucleophilic reactivity of Alkoxide ions toward 2,4-dinitroflyorobenzene and the acidity of alcohols*, Acta Chem. Scand. 18 (1964), pp. 1043–1053.

[264] H. Brink, *Alkylation of alcohols for gas chromatographic analysis by a phase-transfer catalysis technique. Elucidation of reaction mechanism in a one-phase system. Part II*, Acta Pharm. Suecica 17 (1980), pp. 233–248.

[265] S. Takahashi, L.A. Cohen, H.K. Miller, and E.G. Peake, *Calculation of the pK_a values of alcohols from σ^* constants and from the carbonyl frequencies of their esters*, J. Org. Chem. 36 (1971), pp. 1205–1209.

[266] N.S. Marans and R.P. Zelimski, *2,2,2-trinitroethanol: Preparation and properties*, J. Am. Chem. Soc. 72 (1950), pp. 5329–5330.

[267] P.G. Farrell, F.Terrier, and R. Schaal, *The effects of solvation upon the acidities of nitroaromatics*, Tetrahedron Lett. 26 (1985), pp. 2435–2438.

[268] P. Burk and P.V.R. Schleyer, *Why are carboxylic acids stronger acids than alcohols? The electrostatic theory of Siggel-Thomas revisited*, J. Mol. Struct.-Theochem. 505 (2000), pp. 161–167.

[269] P.R. Rablen, *Is the acetate anion stabilized by resonance or electrostatics? A systematic structural comparison*, J. Am. Chem. Soc. 122 (2000), pp. 357–368.

[270] J. Holt and J. M. Karty, *Origin of the acidity enhancement of formic acid over methanol: Resonance versus inductive effects*, J. Am. Chem. Soc. 125 (2003), pp. 2797–2803.

[271] M.R. Siggel and T.D. Thomas, *Why are organic acids stronger acids than organic alcohols?* J. Am. Chem. Soc. 108 (1986), pp. 4360–4363.

[272] M.R. Siggel, T.D. Thomas, and L.J. Saethre, *Ab initio calculation of Brønsted acidities*, J. Am. Chem. Soc. 1988 (1988), pp. 91–96.

[273] M.R. Siggel and A. Streitwieser, Jr., *The role of resonance and inductive effects in the acidity of carboxylic acids*, J. Am. Chem. Soc. 110 (1988), pp. 8022–8028.

[274] K.B. Wiberg, J. Ochterski, and A. Streitwieser, *Origin of the acidity of enols and carboxylic acids*, J. Am. Chem. Soc. 118 (1996), pp. 8291–8299.

[275] G.V. Calder and T.J. Barton, *Actual effects controlling the acidity of carboxylic acids*, J. Chem. Educ. 48 (1971), pp. 338–340.

[276] J.D.S. Goulden, *The OH-vibration frequencies of carboxylic acids and phenols*, Spectrochim. Acta 6 (1954), pp. 129–133.

[277] A.I. Biggs and R.A. Robinson, *The ionisation constants of some substituted anilines and phenols: A test of the Hammett relation*, J. Chem. Soc. 1961 (1961), pp. 388–393.

[278] B.L. Dyatkin, E.P. Mochalina, and I.L. Knunyants, *The acidic properties of fluorine-containing alcohols, hydroxylamines, and oximes*, Tetrahedron 21 (1965), pp. 2991–2995.

[279] R.I. Gelb and J.S. Alper, *Anomalous conductance in electrolyte solutions: A potentiometric and conductometric study of the dissociation of moderately strong acids*, Anal. Chem. 72 (2000), pp. 1322–1327.

[280] L.M. Schwartz, *Ion-pair complexation in moderately strong aqueous acids*, J. Chem. Educ. 72 (1995), pp. 823–826.

[281] U.A. Chaudry and P.L.A. Popelier, *Estimation of pK_a using quantum topological molecular similarity descriptors: Application to carboxylic acids, anilines and phenols*, J. Org. Chem. 69 (2004), pp. 233–241.

[282] G.B. Rocha, R.O. Freire, A.M. Simas, and J.J.P. Stewart, *RM1: A reparameterization of AM1 for H, C, N, O, P, S, F, Cl, Br, and I*, J. Comp. Chem. 27 (2006), pp. 1101–1111.

[283] D.M. Chipman, *Computation of pK_a from dielectric continuum theory*, J. Phys. Chem. A 106 (2002), pp. 7413–7422.

[284] R. Parthasarathi, J. Padmanabhan, M. Elango, K. Chitra, V. Subramanian, and P.K. Chattaraj, *pK_a prediction using group philicity*, J. Phys. Chem. A 110 (2006), pp. 6540–6544.

[285] A.P. Harding, D.C. Wedge, and P.L.A. Popelier, *pK_a prediction from "quantum chemical topology" descriptors*, J. Chem. Inf. Model. 49 (2009), pp. 1914–1924.

[286] A.P. Harding and P.L.A. Popelier, *pK_a Prediction from an ab initio bond length. Part 2. Phenols*, Phys. Chem. Chem. Phys. 13 (2011), pp. 11264–11282.

[287] W.-M. Hoe, A.J. Cohen, and N.C. Handy, *Assessment of a new local exchange functional*, Chem. Phys. Lett. 341 (2001), pp. 319–328.

[288] S. Zhang, *A reliable and efficient first principles-based method for predicting pK$_a$ values. III. Adding explicit water molecules: Can the theoretical slope be reproduced and pK$_a$ values predicted more accurately?* J. Comp. Chem. 33 (2012), pp. 517–526.

[289] F. Eckert, M. Diedenhofen, and A. Klamt, *Towards a first principles prediction of pK$_a$: COSMO-RS and the cluster-continuum approach,* Mol. Phys. 108 (2010), pp. 229–241.

[290] S.K. Burger, S. Liu, and P.W. Ayers, *Practical calculation of molecular acidity with the aid of a reference molecule,* J. Phys. Chem. A 115 (2011), pp. 1293–1304.

[291] J. Sadowski and J. Gasteiger, *From atoms and bonds to three-dimensional atomic coordinates: Automatic model builders,* Chem. Rev. 93 (1993), pp. 2567–2581.

[292] L.D. Freedman and G.O. Doak, *The preparation and properties of phosphonic acids,* Chem. Rev. 57 (1957), pp. 479–523.

[293] A.W. Frank, *The phosphonous acids and their derivatives,* Chem. Rev. 61 (1961), pp. 389–424.

[294] S.S. Patel, *Fosfomycin tromethamine,* Drugs 53 (1997), pp. 637–657.

[295] W. Mabey and T. Mill, *Critical review of hydrolysis of organic compounds in water under environmental conditions,* J. Phys. Chcm. Ref. Data 7 (1978), pp. 383–415.

[296] H.H. Jaffé, L.D. Freedman, and G.O. Doak, *The acid dissociation constants of aromatic phosphonic acids. I. Meta and para substituted compounds,* J. Am. Chem. Soc. 75 (1953), pp. 2209–2211.

[297] J.P. Guthrie, *Tautomerization equilibria for phosphorous acid and its ethyl esters, free energies of formation of phosphorous and phosphonic acids and their ethyl esters, and pK$_a$ values for ionization of the P-H bond in phosphonic acid and phosphonic esters,* Can. J. Chem. 57 (1979), pp. 236–239.

[298] K. Ohta, *Prediction of pK$_a$ values of alkylphosphonic acids,* Bull. Chem. Soc. Jpn. 65 (1992), pp. 2543–2545.

[299] J.J.P. Stewart, *Optimization of parameters for semiempirical methods I. Method,* J. Comp. Chem. 10 (1989), pp. 209–220.

[300] J.J.P. Stewart, *Optimization of parameters for semiempirical methods II. Applications,* J. Comp. Chem. 10 (1989), pp. 221–264.

[301] A. Moser, K. Range, and D.M. York, *Accurate proton affinity and gas-phase basicity values for molecules important in biocatalysis,* J. Phys. Chem. B 114 (2010), pp. 13911–13921.

[302] M.J. Miller, *Syntheses and therapeutic potential of hydroxamic acid based siderophores and analogues,* Chem. Rev. 89 (1989), pp. 1563–1579.

[303] E. Lipczynska-Kochany, *Photochemistry of hydroxamic acids and derivatives,* Chem. Rev. 91 (1991), pp. 477–491.

[304] Y.K. Agrawal and J.P. Shukla, *Para-substituted benzohydroxamic acids: Thermodynamic ionization constants and applicability of Hammett equation,* Aust. J. Chem. 26 (1973), pp. 913–915.

[305] V. Chandrasekhar, B. Ramamoorthy, and S. Nagendran, *Recent developments in the synthesis and structure of organosilanols,* Chem. Rev. 104 (2004), pp. 5847–5910.

[306] N.N. Greenwood and A. Earnshaw, *The Elements*, Pergamon Press, Oxford, 1984.

[307] P.D. Lickiss, *The synthesis and structure of organosilanols*, Adv. Inorg. Chem. 42 (1995), pp. 147–262.

[308] M. Liu, N.T. Tran, A.K. Franz, and J.K. Lee, *Gas-phase acidity studies of dual hydrogen-bonding organic silanols and organocatalysts*, J. Org. Chem. 76 (2011), pp. 7186–7194.

[309] A.G. Schafer, J.M. Wieting, and A.E. Mattson, *Silanediols: A new class of hydrogen bond donor catalysts*, Org. Lett. 13 (2011), pp. 5228–5231.

[310] M.L. Hair and W. Hertl, *Acidity of surface hydroxyl groups*, J. Phys. Chem. 74 (1970), pp. 91–94.

[311] J. Nawrocki, *The silanol group and its role in liquid chromatography*, J. Chromatog. A 779 (1997), pp. 29–71.

[312] A. Méndez, E. Bosch, M. Rosés, and U.D. Neue, *Comparison of the acidity of residual silanol groups in several liquid chromatography columns*, J. Chromatog. A 986 (2003), pp. 33–44.

[313] J.M. Rosenholm, T. Czuryszkiewicz, F. Kleitz, J.B. Rosenholm, and M. Linden, *On the nature of the Brønsted acidic groups on native and functionalized mesoporous siliceous SBA-15 as studied by benzylamine adsorption from solution*, Langmuir 23 (2007), pp. 4315–4323.

[314] R. West and R.H. Baney, *Hydrogen bonding studies. II. The acidity and basicity of silanols compared to alcohols*, J. Am. Chem. Soc. 81 (1959), pp. 6145–6148.

[315] J. Nawrocki, *Silica surface controversies, strong adsorption sites, their blockage and removal. Part I*, Chromatographia, 31 (1991), pp 177–192.

[316] G.B. Cox, *The influence of silica structure on reverse-phase retention*, J. Chromatog. A 656 (1993), pp. 353–367.

[317] S. Ong, X. Zhao, and K.B. Eisenthal, *Polarization of water molecules at a charged interface: Second harmonic studies of the silica/water interface*, Chem. Phys. Lett. 191 (1992), pp. 327–335.

[318] H.-F. Fan, F. Li, R.N. Zare, and K.-C. Lin, *Characterization of two types of silanol groups on fused-silica surfaces using evanescent-wave cavity ring-down spectroscopy*, Anal. Chem. 79 (2007), pp. 3654–3661.

[319] J.R. Rustad, E. Wasserman, A.R. Felmy, and C. Wilke, *Molecular dynamics study of proton binding to silica surfaces*, J. Colloid Interface Sci. 198 (1998), pp. 119–129.

[320] J.A. Tossell, *Calculating the acidity of silanols and related oxyacids in aqueous solution*, Geochim. Cosmochim. Acta 64 (2000), pp. 4097–4113.

[321] F.G. Bordwell, X.-M. Zhang, A.V. Satish, and J.-P. Cheng, *Assessment of the importance of changes in ground-state energies on the bond dissociation enthalpies of the O-H bonds in phenols and the S-H bonds in thiophenols*, J. Am. Chem. Soc. 116 (1994), pp. 6605–6610.

[322] R.M. Borges dos Santos, V.S.F. Muralha, C.F. Correia, R.C. Guedes, B.J. Costa Cabral, and J.A. Martinho Simoes, *S-H bond dissociation enthalpies in thiophenols: A time-resolved photoacoustic calorimetry and quantum chemistry study*, J. Phys. Chem. 106 (2002), pp. 9883–9889.

[323] M.M. Kreevoy, E.T. Harper, R.E. Duvall, H.S. Wilgus, III, and L.T. Ditsch, *Inductive effects on the acid dissociation constants of mercaptans*, J. Am. Chem. Soc. 82 (1960), pp. 4899–4902.

[324] A.K. Chandra, P.-C. Nam, and M.T. Nguyen, *The S-H bond dissociation enthalpies and acidities of para and meta substituted thiophenols: A quantum chemical study*, J. Phys. Chem. A 107 (2003), pp. 9182–9188.

[325] N.E. Hunter and P.G. Seybold, *Theoretical estimation of the aqueous pK_as of thiols*. Mol. Physics (in press 2013).

[326] M. Morgenthaler, E. Schweizer, A. Hoffmann-Roder, F. Benini, R.E. Martin, G. Jaeschke, B. Wagner, H. Fischer, S. Bendels, D. Zimmerli, J. Schneider, F. Diederich, M. Kansy, and K. Muller, *Predicting and tuning physicochemical properties in lead optimization: Amine basicities*, ChemMedChem 2 (2007), pp. 1100–1115.

[327] M.S.B. Munson, *Proton affinities and the methyl inductive effect*, J. Am. Chem. Soc. 87 (1965), pp. 2332–2336.

[328] D.H. Aue, H.M. Webb, and M.T. Bowers, *Quantitative proton affinities, ionization potentials, and hydrogen affinities of alkylamines*, J. Am. Chem. Soc. 98 (1976), pp. 311–317.

[329] E.M. Arnett, F.M.I. Jones, M. Taagepera, W.G. Henderson, J.L. Beauchamp, D. Holtz, and R.W. Taft, *A Complete thermodynamic analysis of the "anomalous order" of amine basicities in solution*, J. Am. Chem. Soc. 94 (1972), pp. 4724–4726.

[330] D.H. Aue, H.M. Webb, and M.T. Bowers, *Quantative relative gas-phase basicities of alkylamines. Correlation with solution basicity*, J. Am. Chem. Soc. 94 (1972): 4726–4728.

[331] F.M.I. Jones and E.M. Arnett, *Thermodynamics of ionization and solution of aliphatic amines in water*, Prog. Phys. Org. Chem. 11 (1974), pp. 263–322.

[332] D.H. Aue, H.M. Webb, and M.T. Bowers, *A thermodynamic analysis of solvation effects on the basicities of alkylamines. An electrostatic analysis of substituent effects*, J. Am. Chem. Soc. 98 (1976), pp. 318–329.

[333] P. Nagy, *Theoretical calculations on the basicity of amines part 2. The influence of hydration*, J. Mol. Struct.-Theochem. 201 (1989), pp. 271–286.

[334] I. Tuñón, E. Silla, and J. Tomasi, *Methylamines basicity calculations. In vacuo and in solution comparative analysis*, J. Phys. Chem. 96 (1992), pp. 9043–9048.

[335] A.D. Headley, *Quantitative analysis of solvation effects and the influence of alkyl substituents on the basicity of amines*, J. Org. Chem. 56 (1991), pp. 3688–3691.

[336] R.C. Rizzo and W.L. Jorgensen, *OPLS all-atom model for amines: Resolution of the amine hydration problem*, J. Am. Chem. Soc. 121 (1999), pp. 4827–4836.

[337] T.C. Bissot, R.W. Parry, and D.H. Campbell, *The physical and chemical properties of the methylhydroxylamines*, J. Am. Chem. Soc. 79 (1957), pp. 796–800.

[338] E.S. Hamborg and G.F. Versteeg, *Dissociation constants and thermodynamic properties of amines an alkanolamines from (293 to 353) K*, J. Chem. Eng. Data 54 (2009), pp. 1318–1328.

[339] H.K. Hall, Jr., *Steric effects on the base strengths of cyclic amines*, J. Am. Chem. Soc. 79 (1957), pp. 5444–5447.

[340] P. Nagy, K. Novak, and G. ASzasz, *Theoretical calculations on the basicity of amines part 1. The use of molecular electrostatic potential for pK_a prediction*, J. Mol. Struct.-Theochem. 201 (1989), pp. 257–270.

[341] J.A. Pople and D.L. Beveridge, *Approximate Molecular Orbital Theory*, McGraw-Hill, New York, 1980.

[342] D.D. Perrin, *Dissociation Constants of Organic Bases in Aqueous Solution*, Supplement no. 1. Butterworths, London, 1972.

[343] S. Zhang, *A reliable and efficient first principles-based method for predicting pK_a values. 4. Organic bases*, J. Comp. Chem. 33 (2012), pp. 2469–2482.

[344] J.S. Murray, T. Brinck, and P. Politzer, *Applications of calculated local surface ionization energies to chemical reactivity*, J. Mol. Struct.-Theochem. 255 (1992), pp. 271–281.

[345] S.H. Hilal, L.A. Carreira, G.L. Baughman, S.W. Karickhoff, and C.M. Melton, *Estimation of ionization constants of azo dyes and related aromatic amines: Environmental implication*, J. Phys. Organic Chem. 7 (1994), pp. 122–141.

[346] I.A. Koppel, J. Koppel, P.-C. Maria, J.-F. Gal, R. Notario, V. Vlasov, and R.W. Taft, *Comparison of the Brönsted acidities of neutral NH-acids in gas phase dimethyl sulfoxide and water*, Int. J. Mass Spectrometry Ion Process. 175 (1998), pp. 61–69.

[347] S. Wold, A. Ruhe, H. Wold, and W.J. Dunn, *The collinearity problem in linear regression—The partial least squares (PLS) approach to generalized inverses*, SIAM J. Sci. Stat. Comput. 5 (1984), pp. 735–743.

[348] J. Catalán, J.L. Abboud, and J. Elguero, *Basicity and acidity of azoles*, Adv. Heterocycl. Chem. 41 (1987), pp. 188–274.

[349] J. Catalán, R.M. Claramunt, J. Elguero, J. Laynez, M. Menéndez, F. Anvia, J.H. Quian, M. Taagepera, and R.W.Taft, *Basicity and acidity of azoles: The annelation effect in azoles*, J. Am. Chem. Soc. 110 (1988), pp. 4105–4111.

[350] J. Catalán, J. Palomar, and J.L.G. de Paz, *On the acidity and basicity of azoles: The Taft scheme for electrostatic proximity effects*, Int. J. Mass Spectrometry Ion Process. 175 (1998), pp. 51–59.

[351] V. Lopez, J. Catalán, R.M. Claramunt, C. Lopez, E. Cayon, and J. Elguero, *On the relationship between thermodynamic pK_a's of azoles and the oxidation potentials of their pentacyanoferrate(II) complexes*, Can. J. Chem. 68 (1990), pp. 958–959.

[352] I.A. Topol, G.J. Tawa, and S.K. Burt, *Calculation of absolute and relative acidities of substituted imidazoles in aqueous solvent*, J. Phys. Chem. A 101 (1997), pp. 10075–10081.

[353] I.-J. Chen and A.D.J. MacKerell, *Computation of the infuence of chemical substitution on the pK_a of pyridine using semiempirical and ab initio methods*, Theor. Chem. Acc. 103 (2000), pp. 483–494.

[354] A.D. Gift, S.M. Stewart, and P.K. Bokashanga, *Experimental determination of pK_a values by use of NMR chemical shifts, revisited*, J. Chem. Educ. 89 (2012), pp. 1458–1460.

[355] M. Sjöström and S. Wold, *A multivariate study of the relationship between the genetic code and the physical-chemical properties of amino acids*, J. Mol. Evol. 22 (1985), pp. 272–277.

[356] G. Bouchoux, *Gas phase basicities of polyfunctional molecules. Part 3. Amino acids*, Mass Spectrom. Rev. 31 (2012), pp. 391–435.

[357] J.T. O'Brian, J.S. Prell, G. Berden, J. Oomens, and E.R. Williams, *Effects of anions on the zwitterion stability of Glu, His and Arg investigated by IRMPD spectroscopy and theory*, Int. J. Mass Spectrometry Ion Process. 297 (2010), pp. 116–123.

[358] A.G. Harrison, *Gas-phase basicities and proton affinities of amino acids and peptides*, Mass Spectrom. Rev. 16 (1997), pp. 201–217.

[359] T.C. Dinadayalane, G.N. Sastry, and J. Leszczynski, *Comprehensive theoretical study towards the accurate proton affinity values of naturally occurring amino acids*, Int. J. Quantum Chem. 106 (2006), pp. 2920–2933.

[360] J. Oomens, J.D. Steill, and B. Redlich, *Gas-phase IR spectroscopy of deprotonated amino acids*, J. Am. Chem. Soc. 131 (2009), pp. 4310–4319.

[361] W. Wang, X. Pu, W. Zheng, N.-B. Wong, and A. Tian, *Some theoretical observations on the 1:1 glycine zwitterion–water complex*, J. Mol. Struct.-Theochem. 626 (2003), pp. 127–132.

[362] J.H. Jensen and M.S. Gordon, *On the number of water molecules necessary to stabilize the glycine zwitterion*, J. Am. Chem. Soc. 117 (1995), pp. 8159–8170.

[363] C.M. Aikens and M.S. Gordon, *Incremental solvation of nonionized and zwitterionic glycine*, J. Am. Chem. Soc. 128 (2006), pp. 12835–12850.

[364] S.M. Bachrach, *Microsolvation of glycine: A DFT study*, J. Phys. Chem. A 112 (2008), pp. 3722–3730.

[365] M.N. Blom, I. Compagnon, N.C. Polfer, G. von Helden, G. Meijer, S. Suhai, B. Paizs, and J. Oomens, *Stepwise solvation of an amino acid: The appearance of zwitterionic structures*, J. Phys. Chem. A 111 (2007), pp. 7309–7316.

[366] S.E. McLain, A.K. Soper, A.E. Terry, and A. Watts, *Structure and hydration of L-proline in aqueous solutions*, J. Phys. Chem. B 111 (2007), pp. 4568–4580.

[367] R. Wu and T.B. McMahon, *Stabilization of zwitterionic structures of amino acids (Gly, Ala, Val, Leu, Ile, Ser and Pro) by ammonium ions in the gas phase*, J. Am. Chem. Soc. 130 (2008), pp. 3065–3078.

[368] A. Parrill, *Amino acid pK$_a$ values*. Michigan State University, http://www.cem.msu.edu/~cem252/sp97/ch24/ch249a.html.

[369] A.E. Fazary, A.F. Mohamed, and N.S. Lebedeva, *Protonation equilibria studies of the statndard α-amino acids in NaNO3 solutions in water and in mixtures of water and dioxane*, J. Chem. Thermodyn. 38 (2006), pp. 1467–1473.

[370] A.E. Fazary, S.E. Ibrahium, and Y.-H. Ju, *Medium effects on the protonation equilibria of L-norvaline*, J. Chem. Eng. Data 54 (2009), pp. 2532–2537.

[371] A. Gero and J.J. Markham, *Studies on pyridines: I. The basicity of pyridine bases*, J. Org. Chem. 16 (1951), pp. 1835–1838.

[372] J.L. Abboud, J. Catalán, J. Elguero, and R.W. Taft, *Polarizability effects on the aqueous solution basicity of substituted pyridines*, J. Org. Chem. 53 (1988), pp. 1137–1140.

[373] H.J. Soscún Machado and A. Hinchliffe, *Relationships between the HOMO energies and pK$_a$ values in monocyclic and bicyclic azines*, J. Mol. Struct.-Theochem. 339 (1995), pp. 255–258.

[374] D.T. Major, A. Laxer, and B. Fischer, *Protonation studies of modified adenine and adenine nucleotides by theoretical calculations and 15-N NMR*, J. Org. Chem. 67 (2002), pp. 790–802.

[375] A. Habibi-Yangjeh, E. Pourbasheer, and M. Danandeh-Jenagharad, *Prediction of basicity constants of various pyridines in aqueous solution using a principal component-genetic algorithm-artificial neural network*, Monatsh. Chem.-Chem. Monthly 139 (2008), pp. 1423–1431.

[376] J.D. Watson and F.H.C. Crick, *Molecular structure of nucleic acids—A structure for deoxyribose nucleic acid*, Nature 171 (1953), pp. 737–738.

[377] J.D. Watson, *The Double Helix: A Personal Account of the Discovery of the Structure of DNA*, Atheneum, New York, 1968.

[378] P.G. Mezey and J.J. Ladik, *A non-empirical molecular orbital study on the relative stabilities of adenine and guanine tautomers*, Theor. Chim. Acta 52 (1979), pp. 129–145.

[379] D. Shugar and K. Szczepaniak, *Tautomerism of pyrimidines and purines in the gas phase and in low-temperature matrices, and some biological implications*, Int. J. Quantum Chem. 20 (1981), pp. 573–581.

[380] A.R. Katritzky and M. Karelson, *AM1 calculations of reaction field effects on the tautomeric equilibria of nucleic acid pyrimidine and purine bases and their 1-methyl analogs*, J. Am. Chem. Soc. 113 (1991), pp. 1561–1566.

[381] A. Broo and A. Holmén, *Ab initio MP2 and DFT calculations of geometry and solution tautomerism of purine and some purine derivatives*, Chem. Phys. 211 (1996), pp. 147–161.

[382] L.M. Salter and G.M. Chaban, *Theoretical study of gas phase tautomerization reactions for the ground and first excited electronic states of adenine*, J. Phys. Chem. A 106 (2002), pp. 4251–4256.

[383] Y.H. Yang, L.C. Sowers, T. Çağin, and W.A. Goddard, I., *First principles calculation of pK_a values for 5-substituted uracils*, J. Phys. Chem. A 105 (2001), pp. 274–280.

[384] S.R. Whittleton, K.C. Hunter, and S.D. Wetmore, *Effects of hydrogen bonding on the acidity of uracil derivatives*, J. Phys. Chem. A 108 (2004), pp. 7709–7718.

[385] I. Pavel, S. Cota, S. Cinta-Pinzaru, and W. Kiefer, *Raman, surface enhanced Raman spectroscopy, and DFT calculations: A powerful approach for the identification and characterization of 5-fluorouracil anticarcinogenic drug species*, J. Phys. Chem. A 109 (2005), pp. 9945–9952.

[386] O. Dolgounitcheva, V.G. Zakrzewski, and J.V. Ortiz, *Ionization energies and Dyson orbitals of thymine and other methylated uracils*, J. Phys. Chem. A 106 (2002), pp. 8411–8416.

[387] M. Di Laudo, S.R. Whittleton, and S.D. Wetmore, *Effects of hydrogen bonding on the acidity of uracil*, J. Phys. Chem. A 107 (2003), pp. 10406–10413.

[388] Y. Huang and H. Kenttämaa, *Theoretical estimations of the 298 K Gasphase acidities of the pyrimidine-based nucleobases uracil, thymine, and cytosine*, J. Phys. Chem. A 107 (2003), pp. 4893–4897.

[389] S.D. Wetmore, R.J. Boyd, and L.A. Eriksson, *A theoretical study of 5-halouracils: Electron affinities, ionization potentials and dissociation of the related anions*, Chem. Phys. Lett. 343 (2004), pp. 151–158.

[390] T.M. Miller, S.T. Arnold, A.A. Viggiano, and A.E.S. Miller, *Acidity of a nucleotide base: Uracil*, J. Phys. Chem. A 108 (2004), pp. 3439–3446.

[391] X. Hu, H. Li, W. Liang, and S. Han, *Systematic study of the tautomerism of uracil induced by proton transfer. Exploration of water stabilization and mutagenicity*, J. Phys. Chem. B 109 (2005), pp. 5935–5944.

[392] M. Mons, I. Dimicoli, F. Piuzzi, B. Tardivel, and M. Elhanine, *Tautomerism of the DNA base guanine and its methylated derivatives as studied by gas-phase infrared and ultraviolet spectroscopy*, J. Phys. Chem. A 106 (2002), pp. 5088–5094.

[393] D.J. Cram, *Fundamentals of Carbanion Chemistry*, Academic Press, New York, 1965.

[394] E. Buncel, E. Buncel, and J.M. Dust, *Carbanion Chemistry: Structures and Mechanisms*, American Chemical Society, Washington, DC, 2003.

[395] A.P. Dicks, *Using hydrocarbon acidities to demonstrate principles of organic structure and bonding*, J. Chem. Educ. 80 (2003), pp. 1322–1327.

[396] J.B. Conant and G.W. Wheland, *The study of extremely weak acids*, J. Am. Chem. Soc. 54 (1932), pp. 1212–1220.

[397] W.K. McEwen, *A further study of extremely weak acids*, J. Am. Chem. Soc. 58 (1936), pp. 1124–1129.

[398] Z.B. Maksić and M. Eckert-Maksić. *A correlation between the acidity and hybridization in non-conjugated hydrocarbons*, Tetrahedron 25 (1969), pp. 5113–5114.

[399] M. Randić and Z.B. Maksić, *Hybridization by the maximum overlap method*, Chem. Rev. 72 (1972), pp. 43–53.

[400] R.G. Pearson and R.L. Dillon, *Rates of ionization of pseudo acids. IV. Relation between rates and equilibria*, J. Am. Chem. Soc. 75 (1953), pp. 2439–2443.

[401] S.S. Novikov, V.I. Slovetskii, S.A. Shevelev, and A.A. Fainzilberg, *Spectrophotometric determination of the dissociation constants of aliphatic nitro compounds*, Russ. Chem. Bull. 4 (1962), pp. 598–605 (Transl. 552–559).

[402] R. Breslow and W. Chu, *Thermodynamic determination of pk_a's of weak hydrocarbon acids using electrochemical reduction data. Triarylmethyl anions, cycloheptatrienyl anion, and triphenyl- and trialkylcyclopropenyl anions*, J. Am. Chem. Soc. 95 (1973), pp. 411–418.

[403] R. Breslow and R. Goodin, *An electrochemical determination of the pK_a of isobutane*, J. Am. Chem. Soc. 98 (1976), pp. 6076–6077.

[404] B.A. Sim, D. Griller, and D.D.M. Wayner, *Reduction potentials for substituted benzyl radicals: pK_a values for the corresponding toluenes*, J. Am. Chem. Soc. 111 (1989), pp. 754–755.

[405] A.J. Kresge, P. Pruszynski, P.J. Stang, and B.L. Williamson, *Base-catalyzed hydrogen exchange and estimates of the acid strength of benzoyl- and (trimethylsilyl)acetylene in aqueous solution. A correlation between acetylene pK, estimates and hydroxide-ion catalytic coefficients for hydrogen exchange*, J. Org. Chem. 56 (1991), pp. 4808–4811.

[406] A. Streitwieser, Jr., J.R. Murdoch, G. Häfelinger, and C.J. Chang, *Acidity of hydrocarbons. XLVI. Equilibrium ion-pair acidities of mono-, di-, and triarylmethanes toward cesium cyclohexylamide*, J. Am. Chem. Soc. 95 (1973), pp. 4248–4254.

[407] M. Meot-Ner, J.F. Liebman, and S.A. Kafafi, *Ionic probes of aromaticity in annelated rings*, J. Am. Chem. Soc. 110 (1988), pp. 5937–5941.

[408] J. Cioslowski, S.T. Mixon, and E.D. Fleischmann, *Electronic structures of trifluoro-, tricyano-, and trinitromethane and their conjugate bases*, J. Am. Chem. Soc. 113 (1991), pp. 4751–4755.

[409] J.S. Murray, T. Brinck, and P. Politzer, *Surface local ionization energies and electrostatic potentials of the conjugate bases of a series of cyclic hydrocarbons in relation to their aqueous acidities*, Int. J. Quantum Chem.: Quantum Biol. Symp. 18 (1991), pp. 91–98.

[410] I. Alkorta and J. Elguero, *Carbon acidity and ring strain: A hybrid HF-DFT approach (Becke3LYP/6–311++G**)*, Tetrahedron 53 (1997), pp. 9741–9748.

[411] J. March, *Advanced Organic Chemistry*, Wiley-Interscience. New York, 1992.

[412] I.A. Topol, G.J. Tawa, R.A. Caldwell, M.A. Eissenstat, and S.K. Burt, *Acidity of organic molecules in the gas phase and in aqueous solvent*, J. Phys. Chem. A 104 (2000), pp. 9619–9624.

[413] I.E. Charif, S.M. Mekelleche, D. Villemin, and N. Mora-Diez, *Correlation of aqueous pK$_a$ values of carbon acids with theoretical descriptors: A DFT study*, J. Mol. Struct.-Theochem. 818 (2007), pp. 1–6.

[414] G.V. Perez and A.L. Perez, *Organic acids without a carboxylic acid functional group*, J. Chem. Educ. 77 (2000), pp. 910–915.

[415] J.P. Richard, G. Williams, and J. Gao, *Experimental and computational determination of the effect of the cyano group on carbon acidity in water*, J. Am. Chem. Soc. 121 (1999), pp. 715–726.

[416] I.A. Koppel, J. Koppel, V. Pihl, I. Leito, M. Mishima, V.M. Vlasov, L.M. Yagupolskii, and R.W. Taft, *Comparison of Brønsted acidities of neutral CH acids in gas phase and dimethyl sulfoxide*, J. Chem. Soc. Perkin Trans. 2 (2000), pp. 1125–1133.

[417] R.W. Taft and F.G. Bordwell, *Structural and solvent effects evaluated from acidities measured in dimethyl sulfoxide and in the gas phase*, Acc. Chem. Res. 21 (1988), pp. 463–469.

[418] A. Kossiakoff and D. Harker, *The calculation of the ionization constants of inorganic oxygen acids from their structures*, J. Am. Chem. Soc. 60 (1938), pp. 2047–2055.

[419] J.E. Ricci, *The aqueous ionization constants of inorganic oxygen acids*, J. Am. Chem. Soc. 70 (1948), pp. 109–113.

[420] P.L. Bayless, *Empirical pKb and pK$_a$ for nonmetal hydrides from periodic table position*, J. Chem. Educ. 60 (1983), pp. 546–547.

[421] F.G. Bordwell, *Equilibrium acidities in dimethyl sulfoxide solution*, Acc. Chem. Res. 21 (1988), pp. 456–463.

[422] C. McCallum and A.D. Pethybridge, *Conductance of acids in dimethylsulphoxide. II. Conductance of some strong acids in DMSO at 25°C*, Electrochim. Acta 20 (1975), pp. 815–818.

[423] L. Pauling, *Why is hydrofluoric acid a weak acid?* J. Chem. Educ. 33 (1956), pp. 16–17.

[424] J.C. McCoubrey, *The acid strength of the hydrogen halides*, Trans. Faraday Soc. 51 (1955), pp. 743–747.

[425] R.T. Myers, *The strength of the hydrohalic acids*, J. Chem. Educ. 53 (1976), pp. 17–19.

[426] L. Pauling, *The strength of hydrohalogenic acids*, J. Chem. Educ. 53 (1976), pp. 762–763.

[427] R.T. Myers, *Hydrogen halides*, J. Chem. Educ. 53 (1976), p. 802.

[428] R. Schmid and A.M. Miah, *The strength of the hydrohalic acids*, J. Chem. Educ. 78 (2001), pp. 116–117.

[429] T.D. Fridgen, *The correlation of binary acid strengths with molecular properties in first-year chemistry*, J. Chem. Educ. 85 (2008), pp. 1220–1221.

[430] E.J. King and G.W. King, *The ionization constant of sulfamic acid from electromotive force measurements*, J. Am. Chem. Soc. 74 (1952), pp. 1212–1215.

[431] E.G.Taylor, R.P. Desch, and A.J. Catotti, *The conductance of sulfamic acid and some sulfamates in water at 25° and conductance measurements of some long chain sulfamates in water and in water–acetone mixtures at 25°*, J. Am. Chem. Soc. 73 (1951), pp. 74–77.

[432] J.J. Klicić, R.A. Friesner, S.-Y. Liu, and W.C. Guida, *Accurate prediction of acidity constants in aqueous solution via density functional theory and self-consistent reaction field methods*, J. Phys. Chem. A 106 (2002), pp. 1327–1335.

[433] E. Raamat, K. Kaupmees, G. Ovsjannikov, A. Trummal, A. Kütt, J. Saame, I. Koppel, I. Kaljurand, L. Lipping, T. Rodima, V. Pihl, I.A. Koppel, and I. Leito, *Acidities of strong neutral Brønsted acids in different media*, J. Phys. Organic Chem. 26 (2013), pp. 162–170.

[434] G.M. Ullmann, *Relations between protonation constants and titration curves in polyprotic acids: A critical view*, J. Phys. Chem. B 107 (2003), pp. 1263–1271.

[435] G. Bouchoux, *Gas-phase basicities of polyfunctional molecules. Part 1. Theory and methods*, Mass Spectrom. Rev. 26 (2007), pp. 775–835.

[436] F. Barni, S.W. Lewis, A. Berti, G.M. Miskelly, and G. Lago, *Forensic application of the luminol reaction as a presumptive test for latent blood detection*, Talanta 72 (2007), pp. 896–913.

[437] J.J. Christensen, D.P. Wrathall, R.M. Izatt, and D.O. Tolman, *Thermodynamics of proton dissociation in dilute aqueous solution. IX. pK, $\Delta H°$, and $\Delta S°$ values for proton ionization from o-, m-, and p-aminobenzoic acids and their methyl esters at 25°*, J. Phys. Chem. 71 (1967), pp. 3001–3006.

[438] B. Noszál and R. Kassai-Tánczos, *Microscopic acid-base equilibria of arginine*, Talanta 38 (1991), pp. 1439–1444.

[439] K. Takács-Novák, A. Avdeef, K.J. Box, B. Podányi, and G. Szász, *Determination of protonation macro- and microconstants and octanol/water partition coefficient of the antiinflammatory drug niflumic acid*, J. Pharm. Biomed. Anal. 12 (1994), pp. 1369–1377.

[440] A. Pagliara, B. Testa, P.-A. Carrupt, P. Jolliet, C. Morin, D. Morin, S. Urien, J.-P. Tillement, and J.-P. Rihoux, *Molecular properties and pharmacokinetic behavior of cetirizine, a zwitterionic H1-receptor antagonist*, J. Med. Chem. 41 (1998), pp. 853–863.

[441] G.A. Olah, G.K.S. Prakash, Á. Molnár, and J. Sommer, *Superacid Chemistry*, 2nd ed., Wiley, Hoboken, NJ, 2009.

[442] N.F. Hall and J.B. Conant, *A study of superacid solutions. I. The use of the chloranil electrode in glacial acetic acid and the strength of certain weak bases*, J. Am. Chem. Soc. 49 (1927), pp. 3047–3061.

[443] R.J. Gillespie, *Fluorosulfuric acid and related superadd media*, Acc. Chem. Res. 1 (1968), pp. 202–209.

[444] R.J. Gillespie and T.E. Peel, *The Hammett acidity function for some superacid systems. II. The systems H2S04-HSO3F, KS03F-HSO3F, HS03F-SO3, HS03F-AsF5, HS03F-SbF5, and HS03F-SbF5-SO3*, J. Am. Chem. Soc. 95 (1973), pp. 5173–5178.

[445] G.A. Olah, *Crossing conventional boundaries in half a century of research*, J. Org. Chem. 70 (2005), pp. 2413–2429.

[446] G.A. Olah, G.K.S. Prakash, and J. Sommer, *Superacids*, Science 206 (1979), pp. 13–20.

[447] G.A. Olah and R.H. Schlosberg, *Chemistry in super acids. I. Hydrogen exchange and polycondensation of methane and alkanes in FS03H-SbF5 ("magic acid") solution. Protonation of alkanes and the intermediacy of CH5+ and related hydrocarbon ions. The high chemical reactivity of "paraffins" in ionic solution reactions*, J. Am. Chem. Soc. 90 (1968), pp. 2726–2727.

[448] A. Kütt, T. Rodima, J. Saame, E. Raamat, V. Maemets, I. Kaljurand, I.A. Koppel, R.Y. Garlyauskayte, Y.L. Yagupolskii, L.M. Yagupolskii, E. Bernhardt, H. Willner, and I. Leito, *Equilibrium acidities of superacids*, J. Org. Chem. 76 (2011), pp. 391–395.

[449] A. Trummal, A. Rummel, E. Lippmaa, I. Koppel, and I.A. Koppel, *Calculations of pK_a of superacids in 1,2-dichloroethane*, J. Phys. Chem. A 115 (2011), pp. 6641–6645.

[450] T. Förster, *Fluoreszenzspektrum und Wasserstoffionenkonzentration*, Naturwissenschaften 36 (1949), pp. 186–187.

[451] A. Weller, *Fast reactions of excited molecules*, Progr. React. Kinet. 1 (1961), pp. 187–214.

[452] M. Rozwadowski, *Effect of pH on the fluorescence of fluorescein solutions*, Acta Phys. Pol. 20 (1961), pp. 1005–1017.

[453] P.G. Seybold, M. Gouterman, and J.B. Callis, *Calorimetric, photometric, and lifetime determinations of fluorescence yields of fluorescein dyes*, Photochem. Photobiol. 9 (1969), pp. 229–242.

[454] E. Vander Donckt, *Acid-base properties of excited states*, Progr. React. Kinet. 5 (1970), pp. 273–299.

[455] S.G. Schulman and J.D. Winefordner, *Influence of pH in fluorescence and phosphorescence spectrometric analysis*, Talanta 17 (1970), pp. 607–616.

[456] S.G. Schulman, *pH effects in molecular luminescence spectrosky*, Rev. Anal. Chem. 1 (1971), pp. 85–111.

[457] J.F. Ireland and P.A.H. Wyatt, *Acid-base properties of electronically excited states of organic molecules*, Adv. Phys. Organic Chem. 12 (1976), pp. 131–221.

[458] T. Förster, *Fluoreszenz Organischer Verbindungen*, Vandenhoek und Ruprecht, Göttingen, 1951.

[459] C.A. Parker, *Photoluminescence of Solutions*, Elsevier, New York, 1968.

[460] N. Agmon, *Elementary steps in excited-state proton transfer*, J. Phys. Chem. A 109 (2005), pp. 13–35.

[461] H.H. Jaffé and H.L. Jones, *Excited state pK values. III. The application of the Hammett equation*, J. Org. Chem. 30 (1965), pp. 964–969.

[462] E.L. Wehry and L.B. Rogers, *Application of linear free energy relations to electronically excited states of monosubstituted phenols*, J. Am. Chem. Soc. 87 (1965), pp. 4234–4238.

[463] L. Stryer, *Excited-state proton transfer reactions: A deuterium isotope effect on fluorescence*, J. Am. Chem. Soc. 88 (1966), pp. 5708–5712.

[464] E.L. Wehry and L.B. Rogers, *Deuterium isotope effects on the protolytic dissociation of organic acids in electronically excited states*, J. Am. Chem. Soc. 88 (1966), pp. 351–354.

[465] E. Vander Donckt and G. Porter, *Acidity constants of aromatic carboxylic acids in the S1 state*, Trans. Faraday Soc. 64 (1968), pp. 3215–3217.

[466] E.L. Wehry and L.B. Rogers, *Variation of excited-state pK_a values with method of measurement*, Spectrochim. Acta 21 (1965), pp. 1976–1978.

[467] B. Marciniak, H. Kozubek, and S. Paszyc, *Estimation of pK_a^* in the first excited singlet state*, J. Chem. Educ. 69 (1992), pp. 247–249.

[468] L.M. Tolbert and K.M. Solntsev, *Excited-state proton transfer: From constrained systems to "super" photoacids to superfast proton transfer*, Acc. Chem. Res. 35 (2002), pp. 19–27.

[469] L.M. Tolbert and J.E. Haubrich, *Enhanced photoacidities of cyanonaphthols*, J. Am. Chem. Soc. 112 (1990), pp. 8163–8165.

[470] L.M. Tolbert and J.E. Haubrich, *Photoexcited proton transfer from enhanced photoacids*, J. Am. Chem. Soc. 116 (1994), pp. 10593–10600.

[471] C.J.T. Grotthuss, *Sur la décomposition de l'eau et des corps qu'elle tient en dissolution à l'aide de l'électricité galvanique*, Ann. Chim. (Paris) 58 (1806), pp. 54–73.

[472] N. Agmon, *The Grotthuss mechanism*, Chem. Phys. Lett. 244:5–6 (1995), pp. 456–462.

[473] D. Marx, *Proton transfer 200 years after von Grotthuss: Insights from ab initio simulations*, Chemphyschem: A Eur. J. Chem. Phys. Phys. Chem. 7 (2006), pp. 1848–70.

[474] S. Cukierman, *Et tu, Grotthuss! and other unfinished stories*, Biochim. Biophys. Acta 1757 (2006), pp. 876–885.

[475] E. Pines and D. Huppert, *pH jump: A relaxational approach*, J. Phys. Chem. 87 (1983), pp. 4471–4478.

[476] K. Adamczyk, M. Prémont-Schwartz, D. Pines, E. Pines, and E.T.J. Nibbering, *Real-time observation of carbonic acid formation in aqueous solution*, Science 326 (2009), pp. 1690–1694.

[477] A. Albert and E.P. Serjeant, *Ionization Constants of Acids and Bases*, Methuen, London, 1962.

[478] R.P. Bell, *The Proton in Chemistry*, 2nd ed., Cornell University Press, Ithaca, NY, 1973.

[479] D. Volgger, A. Zemann, G. Bonn, and M.J. Antal, Jr., *High-speed separation of carboxylic acids by co-electroosmotic capillary electrophoresis with direct and indirect UV detection*, J. Chromatog. A 758 (1997), pp. 263–276.

[480] G.N. Lewis and P.W. Schutz, *The ionization of some weak electrolytes in heavy water*, J. Am. Chem. Soc. 56 (1934), p. 1915.

[481] C.K. Rule and V.K. La Mer, *Dissociation constants of deutero acids by E. M. F. measurements*, J. Am. Chem. Soc. 60 (1938), pp. 1974–1981.

[482] D.C. Martin and J.A.V. Butler, *The dissociation constants of some nitrophenols in deuterium oxide*, J. Chem. Soc. (1939), pp. 1366–1371.

[483] P. Ballinger and F.A. Long, *Acid ionization constants of alcohols. I. Trifluoroethanol in the solvents H_2O and D_2O*, J. Am. Chem. Soc. 81 (1959), pp. 1050–1053.

[484] C.A. Bunton and V.J. Shiner, Jr., *Isotope effects in deuterium oxide solution. I. Acid-base equilibria*, J. Am. Chem. Soc. 83 (1961), pp. 42–47.

[485] R.P. Bell and A.T. Kuhn, *Dissociation constants of some acids in deuterium oxide*, Trans. Faraday Soc. 59 (1963), pp. 1789–1793.

[486] W.P. Jencks and K. Salvesen, *Equilibrium deuterium isotope effects on the ionization of thiol acids*, J. Am. Chem. Soc. 93 (1971), pp. 4433–4436.

[487] E. Bulemela and P.R. Tremaine, *D_2O isotope effects on the ionization of β-naphthol and boric acid at temperatures from 225 to 300°C using UV-visible spectroscopy*, J. Solution Chem. 38 (2009), pp. 805–826.

[488] K.M. Erickson, H. Arcis, D. Raffa, G.H. Zimmerman, and P.R. Tremaine, *Deuterium isotope effects on the ionization constant of acetic acid in H_2O and D_2O by AC conductance from 368 to 548 K at 20 MPa*, J. Phys. Chem. B 115 (2011), pp. 3038–3051.

[489] F.G. Bordwell and D.L. Hughes, *Hammett and Brønsted-type relationships in reactions of 9-substituted fluorenyl anions with benzyl halides*, J. Org. Chem. 45 (1980), pp. 3320–3325.

[490] W.N. Olmstead, Z. Margolin, and F.G. Bordwell, *Acidities of water and simple alcohols in dimethyl sulfoxide solution*, J. Org. Chem. 45 (1980), pp. 3295–3299.

[491] F.G. Bordwell, R.J. McCallum, and W.N. Olmstead, *Acidities and hydrogen bonding of phenols in dimethyl sulfoxide*, J. Org. Chem. 49 (1984), pp. 1424–1427.

[492] R.W. Taft, F. Anvia, M. Taagepera, J. Catalán, and J. Elguero, *Electrostatic proximity effects in the relative basicities and acidities of pyrazole, imidazole, pyridazine, and pyrimidine*, J. Am. Chem. Soc. 108 (1986), pp. 3237–3239.

[493] T. Clark, J.S. Murray, P. Lane, and P. Politzer, *Why are dimethyl sulfoxide and dimethyl sulfone such good solvents?* J. Mol. Modeling 14 (2008), pp. 689–697.

[494] J.R. Pliego, Jr. and J.M. Riveros, *Gibbs energy of solvation of organic ions in aqueous and dimethyl sulfoxide solutions*, Phys. Chem. Chem. Phys. 4 (2002), pp. 1622–1627.

[495] Y. Fu, L. Liu, Y.-M. Wang, J.-N. Li, T.-Q. Yu, and Q.-X. Guo, *Quantum-chemical predictions of redox potentials of organic anions in dimethyl sulfoxide and reevaluation of bond dissociation enthalpies measured by the electrochemical methods*, J. Phys. Chem. A 110 (2006), pp. 5874–5886.

[496] K. Shen, Y. Fu, J.-N. Li, L. Liu, and Q.-X. Guo, *What are the pK_a values of C-H bonds in aromatic heterocyclic compounds in DMSO?* Tetrahedron 63 (2007), pp. 1568–1576.

[497] D. Gao, *Acidities of water and methanol in aqueous solution and DMSO*, J. Chem. Educ. 86 (2009), pp. 864–868.

[498] X.Q. Zhu, C.H. Wang, and H. Liang, *Scales of oxidation potentials,pK$_a$, and BDE of various hydroquinones and catechols in DMSO*, J. Org. Chem. 75 (2010), pp. 7240–7257.

[499] M. Kilpatrick, Jr. and M.L. Kilpatrick, *Relative acid strengths in acetonitrile*, Chem. Rev. 33 (1933), pp. 131–137.

[500] I.M. Kolthoff, S. Bruckenstein, and M.K. Chantooni, Jr., *Acid-base equilibria in acetonitrile. Spectrophotometric and conductometric determination of the dissociation of various acids*, J. Am. Chem. Soc. 83 (1961), pp. 3927–3935.

[501] I.M. Kolthoff and M.K. Chantooni, Jr., *Calibration of the glass electrode in acetonitrile. Shape of potentiometric titration curves. Dissociation constant of picric acid*, J. Am. Chem. Soc. 87 (1965), pp. 4428–4436.

[502] I.M. Kolthoff and M.K. Chantooni, Jr., and S. Bhowmik, *Acid-base properties of mono- and dinitrophenols in acetonitrile*, J. Am. Chem. Soc. 88 (1966), pp. 5430–5439.

[503] J.F. Coetzee and G.R. Padmanabhan, *Dissociation and homoconjugation of certain phenols in acetonitrile*, J. Phys. Chem. 69 (1965), pp. 3193–3196.

[504] A. Kütt, I. Leito, I. Kaljurand, L. Soovali, V.M. Vlasov, L.M. Yagupolskii, and I.A. Koppel, *A comprehensive self-consistent spectrophotometric acidity scale of neutral bronsted acids in acetonitrile*, J. Org. Chem. 71 (2006), pp. 2829–2838.

[505] E.-I. Rõõm, A. Kütt, I. Kaljurand, I. Koppel, I. Leito, I.A. Koppel, M. Mishima, K. Goto, and Y. Miyahara, *Brønsted basicities of diamines in the gas phase, acetonitrile, and tetrahydrofuran*, Chem.: Eur. J. 13 (2007), pp. 7631–7643.

[506] J. Barbosa, J.L. Beltrán, and V. Sanz-Nebot, *Ionization constants of pH reference materials in acetonitrile-water mixtures up to 70% (w/w)*, Anal. Chim. Acta 288 (1994), pp. 271–278.

[507] J. Barbosa, R. Bergés, I. Toro, and V. Sanz-Nebot, *Protonation equilibria of quinolone antibacterials in acetonitrile-water mobile phases used in LC*, Talanta 44 (1997), pp. 1271–1283.

[508] J. Berdys, M. Makowski, M. Makowska, A. Puszko, and L. Chmurzyński, *Experimental and theoretical studies of acid-base equilibria of substituted 4-nitropyridine N-oxides*, J. Phys. Chem. A 107 (2003), pp. 6293–6300.

[509] J.-N. Li, Y. Fu, L. Liu, and Q.-X. Guo, *First-principle predictions of basicity of organic amines and phosphines in acetonitrile*, Tetrahedron 62 (2006), pp. 11801–11813.

[510] S. Şanli, Y. Altun, N. Şanli, G. Alsancak, and J.L. Beltrán, *Solvent effects on pK$_a$ values of some substituted sulfonamides in acetonitrile-water binary mixtures by the UV-spectroscopy method*, J. Chem. Eng. Data 54 (2009), pp. 3014–3021.

[511] M.J. Kamlet, J.L. Abboud, M.H. Abraham, and R.W. Taft, *Linear solvation energy relationships. 23. A comprehensive collection of the solvatochromic parameters, π*, α, and β, and some methods for simplifying the generalized solvatochromic equation*, J. Org. Chem. 48 (1983), pp. 2877–2887.

[512] M. Eckert-Maksić, Z. Glasovac, P. Trošelj, A. Kütt, T. Rodima, I. and Koppel, I.A. Koppel, *Basicity of guanidines with heteroalkyl side chains in acetonitrile*, Eur. J. Org. Chem. 30 (2008), pp. 5176–5184.

[513] F. Eckert, I. Leito, I. Kaljurand, A. Kütt, A. Klamt, and M. Diedenhofen, *Prediction of acidity in acetonitrile solution with COSMO-RS*, J. Comp. Chem. 30 (2009), pp. 799–810.

[514] Z. Glasovac, M. Eckert-Maksić, and Z.B. Maksić, *Basicity of organic bases and superbases in acetonitrile by the polarized continuum model and DFT calculations*, New J. Chem. 33 (2009), pp. 588–597.

[515] R.R. Fraser, M. Bresse, and T.S. Mansour, *pK_a measurements in tetrahydrofuran*, J. Chem. Soc. Chem. Commun. (1983), pp. 620–621.

[516] R.R. Fraser, M. Bresse, and T.S. Mansour, *Ortho lithiation of monosubstituted benzenes: A quantitative determination of pK, values in tetrahydrofuran*, J. Am. Chem. Soc. 105 (1983), pp. 7790–7791.

[517] R.R. Fraser and T.S. Mansour, *Acidity measurements with lithiated amines: Steric reduction and electronic enhancement of acidity*, J. Org. Chem. 49 (1984), pp. 3443–3444.

[518] R.R. Fraser, T.S. Mansour, and S. Savard, *Acidity measurements on pyridines in tetrahydrofuran using lithiated silylamines*, J. Org. Chem. 50 (1985), pp. 3232–3234.

[519] R.R. Fraser, T.S. Mansour, and S. Savard, *Acidity measurements in THF. V. Heteroaromatic compounds containing 5-membered rings*, Can. J. Chem. 63 (1985), pp. 3505–3509.

[520] A. Streitwieser, Jr., D.Z. Wang, M. Stratakis, A. Facchetti, R. Gareyev, A. Abbotto, J.A. Krom, and K.V. Kilway, *Extended lithium ion pair indicator scale in tetrahydrofuran*, Can. J. Chem. 76 (1998), pp. 765–769.

[521] J. Hine and M. Hine, *The relative acidity of water, methanol, and other weak acids in isopropyl alcohol solution*, J. Am. Chem. Soc. 74 (1952), pp. 5266–5271.

[522] M.K. Chantooni, Jr. and I.M. Kolthoff, *Resolution of acid strength in tert-butyl alcohol and isopropyl alcohol of substituted benzoic acids, phenols, and aliphatic carboxylic acids*, Anal. Chem. 51 (1979), pp. 133–140.

[523] I. Leito, T. Rodima, I.A. Koppel, R. Schwesinger, and V.M. Vlasov, *Acid-base equilibria in nonpolar media. 1. A spectrophotometric method for acidity measurements in heptane*, J. Org. Chem. 62 (1997), pp. 8479–8483.

[524] E.-I. Rõõm, I. Kaljurand, I. Leito, T. Rodima, I.A. Koppel, and V.M. Vlasov, *Acid-base equilibria in nonpolar media. 3. Expanding the spectrophotometric acidity scale in heptane*, J. Org. Chem. 68 (2003), pp. 7795–7799.

[525] M.K. Chantooni, Jr. and I.M. Kolthoff, *Comparison of substituent effects on dissociation and conjugation of phenols with those of carboxylic acids in acetonitrile, N,N-dimethylformamide, and dimethyl sulfoxide*, J. Phys. Chem. 80 (1976), pp. 1306–1310.

[526] I.M. Kolthoff, J.J. Lingane, and W.D. Larson, *The relation between equilibrium constants in water and in other solvents*, J. Am. Chem. Soc. 60 (1938), pp. 2512–2515.

[527] M.K. Chantooni, Jr. and I.M. Kolthoff, *Acid-base equilibria in methanol, acetonitrile, and dimethyl sulfoxide in acids and salts of oxalic acid and homologs, fumaric and o-phthalic acids. Transfer activity coefficients of acids and ions*, J. Phys. Chem. 79 (1975), pp. 1176–1182.

[528] A. Streitwieser, Jr., E. Juaristi, and L.L. Nebenzahl, in *Comprehensive Carbanion Chemistry. Part A. Structure and Reactivity*, E. Buncel and T. Durst, eds., Elsevier, Amsterdam, 1980.

[529] I.M. Kolthoff and S. Bruckenstein, *Acid-base equilibria in glacial acetic acid. I. Spectrophotometric determination of acid and base strengths and of some dissociation constants*, J. Am. Chem. Soc. 78 (1956), pp. 1–9.

[530] F.G. Bordwell, J.C. Branca, D.L. Hughes, and W.N. Olmstead, *Equilibria involving organic anions in dimethyl sulfoxide and N-methylpyrrolidin-2-one: Acidities, ion pairing, and hydrogen bonding*, J. Org. Chem. 45 (1980), pp. 3305–3313.

[531] P.A. Giguère, *The great fallacy of the H+ ion and the true nature of H_3O+*, J. Chem. Educ. 56 (1979), pp. 571–575.

[532] A. Kütt, V. Movchun, T. Rodima, T. Dansauer, E.B. Rusanov, I. Leito, I. Kaljurand, J. Koppel, V. Pihl, I. Koppel, G. Ovsjannikov, L. Toom, M. Mishima, M. Medebielle, E. Lork, G.-V. Röschenthaler, I.A. Koppel, and A.A. Kolomeitsev, *Pentakis(trifluoromethyl)phenyl, a sterically crowded and electron-withdrawing group: Synthesis and acidity of pentakis(trifluoromethyl)benzene, -toluene, -phenol, and -aniline*, J. Org. Chem. 73 (2008), pp. 2607–2620.

[533] L.M. Mihichuk, G.W. Driver, and K.E. Johnson, *Brønsted acidity and the medium: Fundamentals with a focus on ionic liquids*, Chemphyschem: Eur. J. Chem. Phys. Phys. Chem. 12 (2011), pp. 1622–1632.

[534] D. Himmel, S.K. Goll, I. Leito, and I. Krossing, *A unified pH scale for all phases*, Angew. Chem. Int. Ed. 49 (2010), pp. 6885–6888.

[535] J.L. Kurz and J.M. Farrar, *The entropies of dissociation of some moderately strong acids*, J. Am. Chem. Soc. 91 (1969), pp. 6057–6062.

[536] Q. Chen, Z. Liu, and C.H. Wong, *An ab initio molecular dynamics study on the solvation of formate ion and formic acid in water*, J. Theor. Comput. Chem. 11 (2012), pp. 1019–1032.

[537] D.D. Perrin, *The effect of temperature on pk values of organic bases*, Aust. J. Chem. 17 (1964), pp. 484–488.

[538] H.S. Harned and R.W. Ehlers, *The dissociation constant of acetic acid from 0 to 60° centigrade*, J. Am. Chem. Soc. 55 (1933), pp. 652–656.

[539] H.S. Harned and R.W. Ehlers, *The dissociation constant of propionic acid from 0 to 60°*, J. Am. Chem. Soc. 55 (1933), pp. 2379–2383.

[540] H.S. Harned and N.D. Embree, *The ionization constant of formic acid from 0 to 60°*, J. Am. Chem. Soc. 56 (1934), pp. 1042–1044.

[541] H.S. Harned and R.O. Sutherland, *The ionization constant of n-butyric acid from 0 to 60°*, J. Am. Chem. Soc. 56 (1934), pp. 2039–2041.

[542] I.M. Klotz, *Parallel change with temperature of water structure and protein behavior*, J. Phys. Chem. 103 (1999), pp. 5910–5916.

[543] D.A. Hinckley, P.G. Seybold, and D.P. Borris, *Solvatochromism and thermochromism of rhodamine solutions*, Spectrochim. Acta 42A (1986), pp. 747–754.

[544] D.A. Hinckley and P.G. Seybold, *Thermodynamics of the rhodamine B lactone-zwitterion equilibrium*, J. Chem. Educ. 64 (1987), pp. 362–364.

[545] D.A. Hinckley and P.G. Seybold, *A spectroscopic/thermodynamic study of the rhodamine B zwitterion equilibrium*, Spectrochim. Acta 44A (1988), pp. 1053–1059.

[546] L.B. Kier, C.-K. Cheng, M. Tute, and P.G. Seybold, *A cellular automata model of acid dissociation*, J. Chem. Inf. Comput. Sci. 38 (1998), pp. 271–275.

[547] S.N. Ayyampalayam, PhD thesis, *Modeling the temperature dependence of pK_a and integration of chemical process models using SPARC.* University of Georgia, Athens, 2004.

[548] W.J. Middleton and R.V. Lindsey, *Hydrogen bonding in fluoro alcohols*, J. Am. Chem. Soc. 86 (1964), pp. 4948–4952.

[549] C.L. Perrin, *Secondary equilibrium isotope effects on acidity*, Adv. Phys. Organic Chem. 44 (2010), pp. 123–171.

[550] R.P. Bell and W.B.T. Miller, *Dissociation constants of formic acid and formic acid-d*, Trans. Faraday Soc. 59 (1963), pp. 1147–1148.

[551] R.P. Bell and J.E. Crooks, *Secondary hydrogen isotope effect in the dissociation constant of formic acid*, Trans. Faraday Soc. 58 (1962), pp. 1409–1411.

[552] C.L. Perrin, B.K. Ohta, J. Kuperman, J. Liberman, and M. Erdélyi, *Stereochemistry of β-deuterium isotope effects on amine basicity*, J. Am. Chem. Soc. 127 (2005), pp. 9641–9647.

[553] C.L. Perrin and Y. Dong, *Secondary deuterium isotope effects on the acidity of carboxylic acids and phenols*, J. Am. Chem. Soc. 129 (2007), pp. 4490–4497.

[554] C.L. Perrin and P. Karri, *Position-specific secondary deuterium isotope effects on basicity of pyridine*, J. Am. Chem. Soc. 132 (2010), pp. 12145–12149.

[555] P.A.M. Dirac, *Quantum mechanics of many-electron systems*, Proc. R. Soc. Lond. A, Contain. Pap. Math. Phys. Character 123 (1929), pp. 714–733.

Index

Printed and bound by CPI Group (UK) Ltd, Croydon, CR0 4YY

21/10/2024

01777105-0020